Augsburger Schriften zur Mathematik, Physik und Informatik
Band 27

Bibliografische Information der Deutschen Nationalbibliothek

Die Deutsche Nationalbibliothek verzeichnet diese Publikation in der Deutschen Nationalbibliografie; detaillierte bibliografische Daten sind im Internet über http://dnb.d-nb.de abrufbar.

ISBN 978-3-8325-3894-1
ISSN 1611-4256

Logos Verlag Berlin GmbH
Comeniushof, Gubener Str. 47,
10243 Berlin
Tel.: +49 030 42 85 10 90
Fax: +49 030 42 85 10 92
INTERNET: http://www.logos-verlag.de

$\mathscr{D}$-MODULES

Local formal convolution of elementary formal meromorphic connections

Dissertation

zur Erlangung des akademischen Grades eines
Doktors der Naturwissenschaften,

vorgelegt der Mathematisch-Naturwissenschaftlichen
Fakultät der Universität Augsburg

vorgelegt von: Robert Gelb

Datum der Abgabe: 01.09.2014

Erstgutachter:	Prof. Dr. Marco Hien
Zweitgutachter:	Prof. Dr. Marc Nieper-Wißkirchen
Tag der mündlichen Prüfung:	28.11.2014

Contents

1. Introduction

According to the classical theorem of Levelt-Turrittin-Malgrange in dimension one, for any meromorphic connection $\mathcal{M}$ over the field $\mathbb{C}((t)) := \mathbb{C}[\![t]\!][t^{-1}]$ of formal Laurent series, there exists a natural number $q \in \mathbb{N}$ such that, after applying the ramification $\pi : u \mapsto u^q = t$, we get the decomposition

$$\pi^+\mathcal{M} \cong \bigoplus_{i=1}^{s} \mathscr{E}^{\varphi_i} \otimes R_i,$$

where $\mathscr{E}^{\varphi_i} := (\mathbb{C}((u)), \mathrm{d} + \mathrm{d}\varphi_i)$ with $\varphi_i \in \mathbb{C}((u))$, and R_i is a finite dimensional $\mathbb{C}((u))$-vector space with regular connection.
In his paper "*An explicit stationary phase formula for the local formal Fourier-Laplace transform*"[1], C. Sabbah uses the direct summands that show up in the above decomposition and takes the direct images under $\rho_i : u \mapsto u^{q_i}$ (in this context, the natural number q from above is defined to be the least common multiple of the q_i), i.e., he considers the so called *elementary formal meromorphic connections*

$$\mathrm{El}(\rho_i, \varphi_i, R_i) := \rho_{i+}\left(\mathscr{E}^{\varphi_i} \otimes R_i\right).$$

Then he shows a similar statement:

Refined Levelt-Turrittin Theorem [1, Cor. 3.3]:
Any finite dimensional $\mathbb{C}((t))$-vector space $\mathcal{M}$ with connection can be written in a unique way as a direct sum

$$\bigoplus_{i} \rho_{i+}\left(\mathscr{E}^{\varphi_i} \otimes R_i\right),$$

in such a way that each $\rho_{i+}\mathscr{E}^{\varphi_i}$ is irreducible and no two $\rho_{i+}\mathscr{E}^{\varphi_i}$ are isomorphic.

□

Conversely one can ask what happens after carrying out specific operations on such elementary formal meromorphic connections. Two results are already well-known:

Tensor product of elementary formal meromorphic connections [1, Prop. 3.8]:

Given two elementary formal meromorphic connections $\mathrm{El}([u \mapsto u^{p_1}], \varphi_1, R_1)$ and $\mathrm{El}([v \mapsto v^{p_2}], \varphi_2, R_2)$, we set $d = \gcd(p_1, p_2), p_i' = p_i/d$, and $\tilde{\rho}_i(\omega) = \omega^{p_i'}$. Then we define

$$\begin{aligned} \rho(\omega) &= \omega^{p_1 p_2/d}, \\ \varphi^{(k)} &= \varphi_1\left(\omega^{p_2'}\right) + \varphi_2\left(\left[e^{2\pi i k d/p_1 p_2}\omega\right]^{p_1'}\right) \quad (k = 0, ..., d-1), \\ R &= \tilde{\rho}_2^+ R_1 \otimes \tilde{\rho}_1^+ R_2. \end{aligned}$$

With this notation,

$$\mathrm{El}\left([u \mapsto u^{p_1}], \varphi_1, R_1\right) \otimes \mathrm{El}\left([v \mapsto v^{p_2}], \varphi_2, R_2\right) \cong \bigoplus_{k=0}^{d-1} \mathrm{El}\left(\rho, \varphi^{(k)}, R\right).$$

□

Another statement (see [1, Th. 5.1]) considers the local formal Laplace transform (also called Fourier transform) $\mathscr{F}_{\pm}^{(0,\infty)}$ with respect to the kernel $e^{\pm\frac{t}{z}}$:

Local formal Fourier transform of elementary formal meromorphic connections:

For any elementary $\mathbb{C}((t))$-vector space $\mathrm{El}(\rho, \varphi, R)$ with irregular connection (i.e., $\varphi \notin \mathbb{C}[\![u]\!]$), the formal Laplace transform $\mathscr{F}_{\pm}^{(0,\infty)}\mathrm{El}(\rho, \varphi, R)$ from 0 to ∞ is the elementary finite dimensional $\mathbb{C}((z))$-vector space with connection $\mathrm{El}\left(\hat{\rho}_{\pm}, \hat{\varphi}, \hat{R}\right)$ with (setting $L_q = \left(\mathbb{C}((u)), \mathrm{d} - \frac{q}{2}\frac{\mathrm{d}u}{u}\right)$)

$$\hat{\rho}_{\pm}(u) = \mp\frac{\rho'(u)}{\varphi'(u)}, \; \hat{\varphi}(u) = \varphi(u) - \frac{\rho(u)}{\rho'(u)}\varphi'(u), \; \hat{R} \cong R \otimes L_q.$$

□

C. Sabbah also shows, based on this formula, how the local formal Fourier transforms $\mathscr{F}_{\pm}^{(s,\infty)}$ (for $s \in \mathbb{C}^{\times}$) and $\mathscr{F}_{\pm}^{(\infty,\infty)}$ can be computed. So, as every $\mathscr{D}$-module can be decomposed with respect to its slopes, this formula is the main step for an alternative computation of the global Fourier transform.

A third operation is the convolution of grade $i \in \mathbb{Z}$ of elementary formal connections $\mathrm{El}(\rho(x), \varphi(x), R)$ and $\mathrm{El}(\eta(y), \psi(y), S)$ with a finite dimensional $\mathbb{C}((x))$- (resp. $\mathbb{C}((y))$-vector space) R (resp. S) with regular connection which is defined by

$$\mathrm{El}(\rho, \varphi, R) *_i \mathrm{El}(\eta, \psi, S) := \mathscr{H}^i \pi_+ \gamma_+ (\mathrm{El}(\rho, \varphi, R) \boxtimes \mathrm{El}(\eta, \psi, S)),$$

where $\pi : \mathbb{C}^2 \to \mathbb{C}, (u,z) \mapsto z$ denotes the projection. Furthermore we consider the following two special settings for γ:

(1) $\gamma : \mathbb{C}^2 \to \mathbb{C}^2, (x,y) \mapsto (x, xy) =: (u,z)$.
In this case, we talk about the *multiplicative convolution*, denoted by $*$.

(2) $\gamma : \mathbb{C}^2 \to \mathbb{C}^2, (x,y) \mapsto (x, x+y) =: (u,z)$.
In this case, we talk about the *additive convolution*, denoted by $*_+$.

Now the question occurs, how the local formal multiplicative (resp. additive) convolution of two elementary connections can be computed and whether the result itself can be represented as a decomposition into a direct sum of elementary connections.

Changing the perspective, we can use a purely formal access in order to study operations on elementary formal meromorphic connections. In his paper "*Microlocalization and stationary phase*"[2], R. G. Lopez gives an important formula for the global Fourier transform F, considered at ∞. This formula is the initial point for our research. We denote by $\mathrm{Sing}(\mathcal{M})$ the set of singularities of $\mathcal{M}$. For $m \in \mathcal{M}$, we put $\hat{m} = 1 \otimes m \in F(\mathcal{M})$.

Formal stationary phase [2, Theorem p. 7]:
Let $\mathcal{M}$ be a holonomic $\mathbb{C}[t]\langle\partial_t\rangle$-module. Then the map

$$\Upsilon : F(\mathcal{M})_\infty \to \bigoplus_{c \in \mathrm{Sing}(\mathcal{M}) \cup \{\infty\}} \mathscr{F}^{(c,\infty)}(\mathcal{M})$$

given by $\Upsilon(\alpha \otimes \hat{m}) = \oplus_c \, \alpha \otimes m$ is an isomorphism of $\mathbb{C}[\![\eta^{-1}]\!][\eta]$-vector spaces with connection. □

Again one can ask whether this formula can be extended to the convolution of two modules $\mathcal{M}$ and $\mathcal{N}$, in the sense that the convolution can be decomposed into a direct sum such that the summands only depend on the singularities of $\mathcal{M}$ and $\mathcal{N}$. Moreover, if such a decomposition is possible, one can study how far it is related to the formulas of local formal convolution from above when we set

$$\mathcal{M} = \mathrm{El}(\rho(x), \varphi(x), R), \; \mathcal{N} = \mathrm{El}(\eta(y), \psi(y), S)$$

with a finite dimensional $\mathbb{C}(\!(x)\!)$- (resp. $\mathbb{C}(\!(y)\!)$-) vector space R (resp. S) with regular connection.

This work starts with an introducing chapter, where some properties of an important tool of $\mathscr{D}$-module theory, namely the functor of moderate nearby cycles, will be introduced.

Chapter 3 starts with a sketch of the main parts of the constructions which we need in order to compute the local formal convolution. In this sketch, we will focus on the action of the nearby cycle functor. Afterwards we will treat the problem of computing the formal germ of the multiplicative convolution of grade zero at the origin, i.e., the aim is a representation of

$$(\mathrm{El}(\mathrm{Id}, \varphi, R) *_0 \mathrm{El}(\mathrm{Id}, \psi, S))_0^\wedge .$$

In fact, this expression can itself be decomposed into a direct sum of holonomic modules - **not** of meromorphic connections because the multiplication map is not proper. Hence we always consider the localized version of the decomposition, denoted by the superscript $(.)^{\mathrm{loc.}}$, which allows us to talk about meromorphic connections, anyway.
The proof of the mentioned local formal Fourier transform formula of C. Sabbah is the key to the proof of such a decomposition because in some degree, we can consider the convolution as a Laplace transform of "higher" grade (which actually is not the same, of course), in the sense that C. Sabbah's proof uses $\psi(y) = -\frac{1}{y}$, and the convolution case treats arbitrary functions $\psi(y)$ of arbitrary finite pole order.

In Chapter 4, we carry out the same computations in the additive case. This means, the goal is a decomposition of

$$(\mathrm{El}(\mathrm{Id}, \varphi, R) *_{+0} \mathrm{El}(\mathrm{Id}, \psi, S))_0^\wedge .$$

The addition map is proper. So the obtained decomposition already consists of meromorphic connections.

The appendix presents an alternative formula for the local formal additive convolution. To be more precise, we will show that there is an isomorphism

$$F(\mathcal{M} *_+ \mathcal{N}) \cong F(\mathcal{M}) \otimes F(\mathcal{N})$$

of meromorphic connections $\mathcal{M}$, $\mathcal{N}$. Due to R. G. Lopez' stationary phase formula, we can give a decomposition of $(\mathcal{M} *_+ \mathcal{N})_s^\wedge$, depending only on the singularities of $\mathcal{M}$ and $\mathcal{N}$. If $\mathcal{M}$ and $\mathcal{N}$ are elementary formal meromorphic connections, this leads to an alternative access to the decomposition problem of Chapter 4, including the choice of arbitrary ramifications ρ and η in the convolution

$$(\mathrm{El}(\rho,\varphi,R) *_{+0} \mathrm{El}(\eta,\psi,S))_s^\wedge, \quad s \in \hat{\mathbb{C}} = \mathbb{C} \cup \{\infty\}.$$

Though the obtained formula is very comfortable (in contrary to the formulas given in the chapters 3 and 4, the alternative formula even computes the regular parts of the decomposition), it is in a purely formal shape. So it does not give rise to the study of Stokes structures.

Acknowledgements

I would like to thank my supervisor Prof. Dr. Marco Hien for providing the topic of this thesis for me. Additionally I am grateful for his mathematical and organisatorical support as well as the helpful discussions. He has always been within my reach and never got tired of our meetings.

Many thanks to Prof. Dr. Marc Nieper-Wißkirchen for his great help in finding a job at university and for the opportunities assisting him in his lectures.

I also want to thank my parents for their support and their faith in me during my time at university.

Darja, thank you for supporting me during all this time. I appreciate your patience.

2. The nearby cycle functor and its properties

In the following chapters, we will use a powerful tool - the nearby cycle functor, denoted by Ψ. For a detailed description of the construction of Ψ see for example [3]. Essential for us are the properties of this functor which we want to describe now:

1) Let X be a smooth complex algebraic variety and $f : X \to \mathbb{C}$ a function on X, defining a reduced divisor $D = f^{-1}(0)$. Let $\mathscr{M}$ be a holonomic left $\mathscr{D}_X$-module such that $\mathscr{M} = \mathcal{O}_X(*D) \otimes_{\mathcal{O}_X} \mathscr{M}$ and the singularities of $\mathscr{M}$ are contained in D. Then the nearby cycles module $\Psi_f \mathscr{M}$ is a holonomic left $\mathscr{D}_X$-module supported on D [1, Section 1.b].

   In short: The nearby cycles module of a holonomic module is holonomic.

2) In [1, Section 1.b], we can also find the following: Let $\pi : X' \to X$ be a proper modification inducing an isomorphism $X' \setminus \pi^{-1}(D) \to X \setminus D$. Now let us set $f' := f \circ \pi, D' := \pi^{-1}(D) := f'^{-1}(0)$. If $\mathscr{M}'$ is a holonomic left $\mathscr{D}_{X'}$-module with $\mathscr{M}' = \mathcal{O}_{X'}(*D') \otimes_{\mathcal{O}_{X'}} \mathscr{M}'$, then as an $\mathcal{O}_X(*D)$-module, it is equal to $\pi_* \mathscr{M}'$. Moreover, as π is proper, we have (see [4, Theorem 4.8.1, p.226])

$$\Psi_f \mathscr{H}^0 \pi_+ \mathscr{M}'(*D) \cong \mathscr{H}^0 \pi|_{D',+} \Psi_{f'} \mathscr{M}', \ \mathscr{H}^j \pi|_{D',+} \Psi_{f'} \mathscr{M}' = 0 \text{ if } j = 0.$$

   In short: The nearby cycle functor commutes with proper direct images.

3) Let $\mathscr{M}$ be a formal holonomic $\mathscr{D}$-module in the variable t. Then, after the classical theorem of Levelt-Turrittin-Malgrange (see page 5), there exists a ramification ρ such that

$$\rho^+ \mathscr{M} \cong \bigoplus_{i=1}^{s} \mathscr{E}^{\varphi_i} \otimes R_i.$$

   Now the nearby cycles functor can be applied to the module $\rho^+(\mathscr{M})[\frac{1}{t}] \otimes \mathscr{E}^{-\alpha(1/t)}$, where $\alpha(1/t) \in t^{-1}\mathbb{C}[t^{-1}]$ is a meromorphic function. This has the following useful effect: $\alpha(1/t)$ appears in in the formal irregular part of $\rho^+(\mathscr{M})$ if and only if $\Psi_t\left(\rho^+(\mathscr{M})[\frac{1}{t}] \otimes \mathscr{E}^{-\alpha(1/t)}\right) \neq 0$ (cf. [5, Example 5.2.1]). Moreover one can compute important data of the corresponding regular part as for example its rank is given by the number $\dim_{\mathbb{C}}\left(\Psi_t\left(\rho^+(\mathscr{M})[\frac{1}{t}]\otimes\mathscr{E}^{-\alpha(1/t)}\right)_0\right)$. [6] and [7] describe the usage of this theory in the case of two variables.

In short: The nearby cycle functor gives a criterion for a meromorphic function to show up in the local formal decomposition of a $\mathscr{D}$-module. Moreover facts about the corresponding regular part can be computed.

4) Let (x, y) be coordinates on $\mathbb{C}^2$. Let $\mathscr{M}$ be a regular holonomic $\mathscr{D}_{\mathbb{C}^2}$-module. We consider the module $\Psi_x\mathscr{P}$ with $\mathscr{P} = \mathscr{M}[\frac{1}{xy}]\otimes\mathscr{E}^{g(x,y)}$ and a rational function g. Now it can happen that g has singularities. In the cases considered in this work, g has just a singularity at $(0,0)$. It can be resolved with the help of a finite composition $\nu : Y \to \mathbb{C}^2$ of blowups of the point $(0,0)$. One cannot apply the commutation property of Ψ and ν_+ directly (ν is not proper in general). [7, Lemma 4.5] and [8, Proposition 8.16] allow us to install a commutation property with the help of slight modifications:

$$\Psi_x\mathscr{P} \cong \Psi_x\left(\nu_+\nu^+\mathscr{P}\right) \cong \tilde{\nu}_+\Psi_{x\circ\nu}\left(\iota_+\nu^+\mathscr{P}\right)$$

with a convenient embedding ι and the induced map $\tilde{\nu} : Y\times\{0\} \to \mathbb{C}^2\times\{0\}$ (see Lemma 3.3.2.6).

In short: The nearby cycle functor commutes with direct images under blowups of points.

5) Special nearby cycles modules can be computed explicitly. We use [1, Section 1.b] again: Let $X = \mathbb{C}^2$ with coordinates (x, y) and $f(x, y) := x^{m_1}y^{m_2}$, where $m_1 \in \mathbb{N}_0$ and $m_2 \in \mathbb{N}$. Set $D_1 := \{y = 0\}$ with coordinate x and $D := \{x^{m_1}y^{m_2} = 0\}$. Let $\mathscr{R}$ be a local free $\mathcal{O}_X$-module of finite rank with a flat connection having regular singularities along D. So $\mathscr{R}$ is also a regular holonomic $\mathscr{D}_X$-module. The $\mathscr{D}_X$-module $\Psi_f\mathscr{R}$ is supported on D. Moreover, if $m_1 \neq 0$, $(\Psi_f\mathscr{R})[x^{-1}]$ is supported on D_1 and is the direct image (in the sense of left $\mathscr{D}_X$-modules) by the inclusion $D_1 \hookrightarrow \mathbb{C}^2$ of a regular holonomic $\mathscr{D}_{D_1}$-module localized and smooth away from $\{x = 0\}$. Due to Kashiwara's equivalence, we do not have to differ between both.

2.1 Proposition: With the previous setting, for any $\lambda \in \mathbb{C}^*$,

a) $(n_1, n_2) \in \mathbb{N}^2 \Rightarrow \Psi_f\left(\mathscr{E}^{\lambda/x^{n_1}y^{n_2}}\otimes\mathscr{R}\right) = 0$.

b) $n_1 \in \mathbb{N} \Rightarrow \Psi_f\left(\mathscr{E}^{\lambda/x^{n_1}}\otimes\mathscr{R}\right)$ is supported on D_1; it is isomorphic to $\mathscr{E}^{\lambda/x^{n_1}}\otimes\left((\Psi_f\mathscr{R})\,[x^{-1}]\right)$. □

We also want to refer to the Lemmas 3.5.1 - 3.5.3 in [9] and Section 7 in [7].

3. The multiplicative convolution of elementary mer. connections

The goal of this chapter is the local formal description of the multiplicative convolution of elementary formal meromorphic connections, introduced by C. Sabbah [1, Def. 2.1]:

3.1 Definition: Given $\rho \in u\mathbb{C}[\![u]\!], \varphi \in \mathbb{C}(\!(u)\!)$ and a finite dimensional $\mathbb{C}(\!(u)\!)$-vector space R with regular connection ∇, we define the associated elementary finite dimensional $\mathbb{C}(\!(t)\!)$-vector space with connection (the *elementary formal meromorphic connection*) by

$$\mathrm{El}(\rho, \varphi, R) := \rho_+(\mathscr{E}^\varphi \otimes R) \text{ with } \mathscr{E}^\varphi \cong \mathscr{E}^\psi \Leftrightarrow \varphi \equiv \psi \mod \mathbb{C}[\![u]\!].$$

3.2 Remark: 1) As

$$\mathscr{E}^\varphi \cong \mathscr{E}^\psi \Leftrightarrow \varphi \equiv \psi \mod \mathbb{C}[\![u]\!],$$

we can assume $\varphi \in u^{-1}\mathbb{C}[u^{-1}]$ without loss of generality.

2) Any elementary formal meromorphic connection $\mathrm{El}(\rho, \varphi, R)$ is defined over the field of convergent power series $\mathbb{C}(\{t\})$ [1, p. 4]. But we can extend $\mathrm{El}(\rho, \varphi, R)$ to a meromorphic connection on $\mathbb{P}^1$ via the representation

$$\mathrm{El}(\rho, \varphi, R) = (\rho_+\mathscr{E}^\varphi) \otimes \tilde{R}.$$

This connection has a regular singularity at ∞. $\tilde{R}$ denotes a finite dimensional $\mathbb{C}(\!(t)\!)$-vector space with regular connection such that $\rho^+\tilde{R} = R$.

3.3 Definition: Let $\mathrm{El}(\rho, \varphi, R)$, $\mathrm{El}(\eta, \psi, S)$ be elementary formal meromorphic connections, where $\rho \in x\mathbb{C}[\![x]\!], \varphi \in \mathbb{C}(\!(x)\!), \eta \in y\mathbb{C}[\![y]\!], \psi \in \mathbb{C}(\!(y)\!)$. Then the *multiplicative convolution* $\mathrm{El}(\rho, \varphi, R) *_i \mathrm{El}(\eta, \psi, S)$ of degree $i \in \mathbb{Z}$ is defined as

$$\mathscr{H}^i\pi_+\gamma_+\left(\left((\rho_+\mathscr{E}^\varphi) \otimes \tilde{R}\right) \boxtimes \left((\eta_+\mathscr{E}^\psi) \otimes \tilde{S}\right)\right),$$

where $\gamma : \mathbb{C}^2 \to \mathbb{C}^2, (x, y) \mapsto (x, xy) =: (u, z)$ and $\pi : \mathbb{C}^2 \to \mathbb{C}, (u, z) \mapsto z$.

Let us concentrate on the convolution of degree 0, i.e., $\mathrm{El}(\rho,\varphi,R) *_0 \mathrm{El}(\eta,\psi,S)$. This is is the only interesting cohomology module after C. R [6, Example 0.4].
In order to simplify the direct image ρ in the representation of an elementary formal meromorphic connection $\mathrm{El}(\rho,\varphi,R)$, we use the proof of [1, Lemma 2.2]: Let ρ be of the form $\sum_{i\geq p} a_i u^i$ with $a_p \neq 0$. So $\rho = u^p \sum_{i\geq 0} b_i u^i$ with $p \geq 1, b_i := a_{p+i}$. Then there exists $\lambda \in u\mathbb{C}[\![u]\!]$ with $\lambda'(0) \neq 0$ and $\rho(\lambda(u)) = u^p$. This gives

$$\begin{aligned}\mathrm{El}(\rho,\varphi,R) &= \rho_+(\mathscr{E}^\varphi \otimes R) = \left(\rho\circ\lambda\circ\lambda^{-1}\right)_+(\mathscr{E}^\varphi \otimes R) \cong (\rho\circ\lambda)_+\left(\left(\lambda^{-1}\right)_+(\mathscr{E}^\varphi \otimes R)\right)\\ &= (\rho\circ\lambda)_+\left(\lambda^+(\mathscr{E}^\varphi \otimes R)\right) \cong (\rho\circ\lambda)_+\left(\mathscr{E}^{\varphi\circ\lambda} \otimes R\right) = \mathrm{El}([u\mapsto u^p],\varphi\circ\lambda,R).\end{aligned}$$

So, without loss of generality, we can assume that $\rho(x) = x^p, \eta(y) = y^q$.

3.4 Remark: As γ is not proper, we can only decompose the multiplicative convolution into a direct sum of holonomic modules. As the modules $\mathrm{El}(\rho,\varphi,R)$ are defined as meromorphic connections, we give the localized version of the decomposition, denoted by $(.)^{\mathrm{loc.}}$.

3.1. Nearby cycle functor and convolution of elementary formal meromorphic connections

We give a sketch of the following procedure, where we want to compute the formal decomposition of the multiplicative convolution of the meromorphic connections $\mathscr{E}^\varphi \otimes R$ and $\mathscr{E}^\psi \otimes S$ at the origin, where $\varphi \in \mathbb{C}[\![x]\!][x^{-1}] =: \mathbb{C}(\!(x)\!)$ (resp. $\psi \in \mathbb{C}(\!(y)\!)$) is a meromorphic function with pole order n (resp. m) and R (resp. S) a finite dimensional $\mathbb{C}(\!(x)\!)$- (resp. $\mathbb{C}(\!(y)\!)$-)vector space with regular connection. We will also use the sketched method in Chapter 4, where we treat the case of additive convolution.
The multiplicative convolution of the given meromorphic connections is defined as

$$(\mathscr{E}^\varphi \otimes R) *_0 (\mathscr{E}^\psi \otimes S) := \mathscr{H}^0\pi_+\gamma_+((\mathscr{E}^\varphi \otimes R) \boxtimes (\mathscr{E}^\psi \otimes S))$$

with $\gamma : (x,y) \mapsto (x,xy) =: (u,z)$ and $\pi : (u,z) \mapsto z$. If ψ has pole order 1, the computation (Section 3.2) is just similar to the proof of the local formal Laplace transform given under [1, Theorem 5.1]. If ψ has pole order $m > 1$, we have to take several steps:

i) We give parametrizations of the singular locus of the convolution (see Section 3.3.1). Roughly speaking, these are given by pairs $(\rho(t), \alpha_k(t) + \delta_k(t))$, $k = 1,...,r$, where $r \leq m$ denotes the number of irreducible components of the polynomial which describes the curve that has to be parameterized (see Lemma 3.3.2.1). Moreover it is $\rho(t) = t^p$ (with $p \in \mathbb{N}$), $\alpha_k(t) \in t^{-1}\mathbb{C}[t^{-1}]$ and $\delta_k(t) \in \mathbb{C}\{t\}$ for all k.

ii) Next we have to cancel redundant pairs. A pair $(\rho(t), \alpha_k(t) + \delta_k(t)), k \in \{1, ..., r\}$, is called redundant if there is another pair $(\rho(t), \alpha_l(t) + \delta_l(t)), l \in \{1, ..., r\} \setminus \{k\}$, and a p-th root of unity ζ such that

$$\alpha_k(\zeta t) + \delta_k(\zeta t) = \alpha_l(t) + \delta_l(t).$$

After a suitable renumbering, we receive a minimal set of parametrizations

$$\{(\rho(t), \alpha_k(t)) | k = 1, ..., r^*\}$$

with $r^* \leq r$. Note that only the meromorphic parts are needed.

iii) We use C. Roucairols main result (see Th. 3.3.2.4 on page 31 resp. [6, Th. 0.1]) in order to give a first decomposition of the convolution. This means, after applying the ramification ρ, we have

$$\rho^+((\mathscr{E}^\varphi \otimes R) *_0 (\mathscr{E}^\psi \otimes S))_0^\wedge \cong \bigoplus_{\alpha \in \Gamma} \mathscr{E}^\alpha \otimes T_\alpha,$$

where $\Gamma \subset t^{-1}[t^{-1}]$ is the finite set $\{\alpha_k(\xi t) | \xi^p = 1, k = 1, ..., r^*\}$.

iv) The decomposition under iii) is not the desired one. We want to prove a decomposition of the form

$$((\mathscr{E}^\varphi \otimes R) *_0 (\mathscr{E}^\psi \otimes S))_0^\wedge \cong \bigoplus_{i=1}^{h} \omega_{i,+}(\mathscr{E}^{\sigma_i} \otimes T_i)$$

with $h \leq r^*$ (note that there still can be functions $\alpha_l(t)$ and $\alpha_k(t)$ with $k \neq l$ and a ξ with $\xi^p = 1$ and $\alpha_k(\xi t) = \alpha_l(t)$ because we omitted the holomorphic parts $\delta_j(t)$), $\sigma_i \in \Gamma \subset t^{-1}\mathbb{C}[t^{-1}]$ and $\omega_i = \rho$ for all i. To do this, we choose a meromorphic function $\alpha \in \Gamma$ and compute the nearby cycles module (with the new variable η)

$$\Psi_\eta \left(\mathscr{E}^{\varphi(u) + \psi\left(\frac{\rho(\eta)}{u}\right) - \alpha(\eta)} \otimes (R \boxtimes S) \right).$$

The detailed proof will show that for this step we need the fact that Ψ commutes with proper direct images.

v) The function $\varphi(u)+\psi\left(\frac{\rho(\eta)}{u}\right)-\alpha(\eta)$ has a singularity at $(u,\eta)=(0,0)$, so we use a finite sequence of blowups in order to solve it. The fact that Ψ commutes with direct images under blowups (we will make this property more explicit later) and the fact that the support of the module is contained in exactly one chart of the blowing-up space (Lemma 3.3.2.6), given by $e_0:(v,\eta)\mapsto\left(v\eta^l,\eta\right)$ for a certain $l\in\mathbb{N}$ (this can be shown with Property 5 in Chapter 2), reduces our proof to the computation of a module of the form

$$\Psi_\eta\left(\mathscr{E}^{g(v,\eta)/v^{n_1}\eta^{n_2}}\otimes e_0^+(R\boxtimes S)\right). \tag{3.1}$$

Applying Property 5 of Chapter 2 gives that this module is supported at most on the set $\{g(v,0)=0\}$. [1, Lemma 5.5] states that (3.1) is not zero iff values $\bar{v}\in\mathbb{C}$ exist such that $(v-\bar{v})^2$ divides $g(v,\eta)$. We will show that for every ξ with $\xi^p=1$, the corresponding functions $g_{\alpha(\xi t)}$ have the same number of values with $(v-\bar{v})^2|g_{\alpha(\xi t)}(v,\eta)$ (Proposition 3.3.2.7) and that for every $\alpha\in\Gamma$ there exists at least one such $\bar{v}$ (Lemma 3.3.2.9).
But one can show even more: Let $\Gamma=\biguplus_{i=1}^{h}\Gamma_i, h\leq r^*$, be a disjoint decomposition of Γ in subsets

$$\Gamma_i:=\{\alpha(t)\in\Gamma\,|\,\exists\xi\in\mathbb{C}^*:\xi^p=1\wedge\alpha(\xi t)=\alpha_i(t)\}\,.$$

Then the regular parts T_{α_i} in the decomposition under iii) are isomorphic for all $\alpha_i\in\Gamma_i$ (Corollary 3.3.2.8). Hence we can write this decompostion in the form

$$((\mathscr{E}^\varphi\otimes R)*_0(\mathscr{E}^\psi\otimes S))_0^\wedge\cong\bigoplus_{i=1}^{h}\rho_+(\mathscr{E}^{\alpha_i}\otimes T_i).$$

In the proof, we cannot determine the number h of direct summands. But it is easy to determine h in any concrete computation because a disjoint decomposition of Γ can always be found.

3.2. A special case

This section is devoted to the computation of the formalization of the convolution $\mathrm{El}(\mathrm{Id},\varphi,R)*_0\mathrm{El}\left(\mathrm{Id},-\frac{1}{y},\mathbf{1}\right)$ in the neighbourhood of the origin, where by $\mathbf{1}$ we denote the field $\mathbb{C}((y))$ together with its natural connection d. Let Δ be a disc centered at the origin of $\mathbb{C}$ and let $\mathbb{P}^1$ be the projective line with affine chart $\mathbb{C}$. In order to study the formalization $\mathrm{El}(\mathrm{Id},\varphi,R) *_0 \mathrm{El}\left(\mathrm{Id},-\frac{1}{y},\mathbf{1}\right)$ at the origin of Δ, we want to bring the module in the form of [1, Theorem 4.3] at first: We have

$$\begin{aligned}
\mathrm{El}(\mathrm{Id},\varphi,R) *_0 \mathrm{El}\left(\mathrm{Id},-\frac{1}{y},\mathbf{1}\right) &= \mathscr{H}^0\pi_+\gamma_+\left(\left(\mathscr{E}^{\varphi(x)}\otimes R\right)\boxtimes\mathscr{E}^{-1/y}\right)\\
&\cong \mathscr{H}^0\pi_+\gamma_+\left(\mathscr{E}^{\varphi(x)-1/y}\otimes\left(R\boxtimes\mathbb{C}(\!(y)\!)\right)\right)\\
&= \mathscr{H}^0\pi_+\gamma_+\left(\mathscr{E}^{\varphi(x)-1/y}\otimes R(\!(y)\!)\right)\\
&\cong \mathscr{H}^0(\pi\circ\gamma)_+\left(\mathscr{E}^{\varphi(x)-1/y}\otimes T\right),
\end{aligned}$$

where the first isomorphism holds because $\mathrm{El}(\rho,\varphi,R)$ only depends on the coordinate x and $\mathrm{El}\left(\mathrm{Id},-\frac{1}{y},\mathbf{1}\right)$ only depends on the coordinate y, and $T := R(\!(y)\!)$ is a finite dimensional $\mathbb{C}(\!(x,y)\!)$-vector space with regular connection. The second isomorphism is a consequence of [10, Th. 2.2.5 (1)] resp. [10, Rem. 2.2.9]
The next step will be the transform of the function $f(x,y) := \pi\circ\gamma : U \to \mathbb{C}$ into the projection $p_1 : \Delta\times\mathbb{P}^1 \to \Delta$, and of $g(x,y) := \varphi(x) - \frac{1}{y} : U \to \mathbb{C}$ into the projection $p_2 : \Delta\times\mathbb{P}^1 \to \mathbb{P}^1$ (where U is a neighbourhood of $(0,0)$). In her paper [6, p. 3], C. Roucairol uses the diagram

$$U \xrightarrow{(f,g)} \mathbb{C}\times\mathbb{C} \overset{i}{\hookrightarrow} \mathbb{P}^1\times\mathbb{P}^1 \overset{i_0}{\hookleftarrow} \Delta\times\mathbb{P}^1$$

and considers $\mathscr{M} := i_0^*(i_+(f,g)_+T)^{\mathrm{an}}$ which consists of a complex of $\mathscr{D}_{\Delta\times\mathbb{P}^1}$-modules with regular holonomic cohomology modules $\mathscr{H}^k\mathscr{M}$. Now we can apply the next theorem to the given module $\mathscr{H}^0 f_+\left(\mathscr{E}^g\otimes T\right)$.

3.2.1 Theorem [6, Th. 0.3]: At the origin of Δ, the two modules $\mathscr{H}^0 f_+(\mathscr{E}^g\otimes T)$ and $\mathscr{H}^0 p_{1+}(\mathscr{E}^{p_2}\otimes\mathscr{H}^0\mathscr{M})$ have the same formal irregular parts. □

So, with [6, Introduction], $\mathscr{H}^0 p_{1+}(\mathscr{E}^{p_2}\otimes\mathscr{H}^0\mathscr{M})$ (and $\mathscr{H}^0 f_+(\mathscr{E}^g\otimes T)$ [6, Ex. 0.4]) is the only interesting cohomology module. But in order to apply [1, Th. 4.3] for an explicit computation of $\mathscr{H}^0 f_+(\mathscr{E}^g\otimes T)$, we have to face several problems:

a) Assumptions [1, 4.1] and [1, 4.2] have to be satisfied which is not easy to show in general.

b) As we have mentioned before, just the formal irregular parts of $\mathscr{H}^0 f_+(\mathscr{E}^g\otimes T)$ and $\mathscr{H}^0 p_{1+}(\mathscr{E}^{p_2}\otimes\mathscr{H}^0\mathscr{M})$ are identical.

To avoid these problems, we use the proof of the main result of [1], where C. Sabbah introduces a formula for the computation of the local formal Laplace transform of elementary formal meromorphic connections. We want to modify his proof in such a way that we obtain a decomposition formula for $\left(\mathrm{El}(\mathrm{Id}, \varphi, R) *_0 \mathrm{El}\left(\mathrm{Id}, -\frac{1}{y}, \mathbf{1}\right)\right)_0^{\wedge,\mathrm{loc.}}$.

3.2.2 Theorem: Let $\mathrm{El}(\mathrm{Id}, \varphi, R)$ be a finite dimensional $\mathbb{C}((x))$-vector space with irregular connection. Then $\left(\mathrm{El}(\mathrm{Id}, \varphi, R) *_0 \mathrm{El}\left(\mathrm{Id}, -\frac{1}{y}, \mathbf{1}\right)\right)_0^{\wedge,\mathrm{loc.}}$ is the $\mathbb{C}((z))$-vector space with connection $\mathrm{El}(\hat{\rho}, \hat{\varphi}, \hat{R})$ with (setting $L_n = \left(\mathbb{C}((u)), \mathrm{d} - \frac{n}{2}\frac{\mathrm{d}u}{u}\right)$)

$$\hat{\rho}(u) = \frac{1}{\varphi'(u)}, \ \hat{\varphi}(u) = \varphi(u) - u\varphi'(u), \ \hat{R} \cong R \otimes L_n$$

Proof (cf. [1, Proof of Th. 5.1]): Choosing an algebraic model for $\mathrm{El}(\mathrm{Id}, \varphi, R)$, we can assume that $\varphi(x) = x^{-n}a(x)$ where $a \in \mathbb{C}[x]$ has degree $< n$ and $a(0) \neq 0$; we moreover assume that R is a free $\mathbb{C}[x, x^{-1}]$-module with a connection having a regular singularity at $x = 0$ and $x = \infty$ and no other pole. Above computations showed that

$$\mathrm{El}(\mathrm{Id}, \varphi, R) *_0 \mathrm{El}\left(\mathrm{Id}, -\frac{1}{y}, \mathbf{1}\right) \cong \mathscr{H}^0\pi_+\gamma_+\left(\mathscr{E}^{\varphi(x)-1/y} \otimes R[y, y^{-1}]\right).$$

Computing the direct image of $\mathscr{E}^{\varphi(x)-1/y} \otimes R[y, y^{-1}]$ under γ can be done immediately. We have

$$\mathscr{H}^0\pi_+\gamma_+\left(\mathscr{E}^{\varphi(x)-1/y} \otimes R[y, y^{-1}]\right) \cong \mathscr{H}^0\pi_+\left(\mathscr{E}^{\varphi(u)-u/z} \otimes R[z, z^{-1}]\right).$$

Let $\mathbb{G}_{m,z}$ denote the torus in the coordinate z and let the morphism

$$\kappa : \mathbb{G}_{m,z} \times \mathbb{G}_{m,u} \to \mathbb{G}_{m,z} \times \mathbb{A}^1_v, (z, u) \mapsto \left(z, \varphi(u) - \frac{u}{z}\right)$$

be given. Then the same heuristic considerations as in the proof of [1, Th. 5.1] give that the singular locus of $\kappa_+(R[z, z^{-1}])$ is the curve parameterized by

$$\mathbb{G}_{m,u} \ni u \mapsto (\hat{\rho}(u), \hat{\varphi}(u))$$

with $\hat{\rho}(u) := \frac{1}{\varphi'(u)}$ and $\hat{\varphi}(u) := \varphi(u) - u\varphi'(u)$.
We apply the ramification $\hat{\rho} : \eta \mapsto z$ to $\mathscr{E}^{\varphi(u)-u/z} \otimes R[z, z^{-1}]$ and then take the direct image by the projection to $\mathbb{G}_{m,\eta}$, i.e., we consider

$$\phi_+\left(\mathscr{E}^{\varphi(u)-\frac{u}{\hat{\rho}(\eta)}} \otimes R[\eta, \eta^{-1}]\right)$$

with $\phi : \mathbb{G}_{m,u} \times \mathbb{G}_{m,\eta} \to \mathbb{G}_{m,\eta}, (u, \eta) \mapsto \eta$. In order to compute the regular part with respect to the exponential term $\mathscr{E}^{\hat{\varphi}(\eta)}$, we use the same method as in C. Sabbahs proof of [1, Th. 4.3]: we tensor this direct image by $\mathscr{E}^{-\hat{\varphi}(\eta)}$ and then take the moderate nearby cycles functor Ψ_η of the resulting meromorphic connection:

$$\Psi_\eta\left(\phi_+\left(\mathscr{E}^{\varphi(u)-\frac{u}{\hat{\rho}(\eta)}}\otimes R[\eta,\eta^{-1}]\right)\otimes\mathscr{E}^{-\hat{\varphi}(\eta)}\right)$$

Due to [4, Th. 4.8-1], the nearby cycles functor and proper direct images commute. So the above module is isomorphic to

$$\phi_+\Psi_{\eta\circ\phi}\left(\left(\mathscr{E}^{\varphi(u)-\frac{u}{\hat{\rho}(\eta)}}\otimes R[\eta,\eta^{-1}]\right)\otimes\phi^+\mathscr{E}^{-\hat{\varphi}(\eta)}\right)$$

But ϕ is the canonical projection and $\phi^+\mathscr{E}^{-\hat{\varphi}(\eta)}\cong\mathscr{E}^{-\hat{\varphi}(\eta)}$. This means, the problem reduces to the computation of

$$\Psi_\eta\left(\mathscr{E}^{\varphi(u)-\hat{\varphi}(\eta)-u/\hat{\rho}(\eta)}\otimes R[\eta,\eta^{-1}]\right).$$

In the following, we will use the notation we have introduced at the beginning of the proof: $\varphi(u)=u^{-n}a(u), u\varphi'(u)=u^{-n}b(u)$ with $b(u)=-na(u)+ua'(u)$ and $a(u)=\sum_{k=0}^{n-1}a_ku^k$. Hence

$$b(u)=-n\sum_{k=0}^{n-1}a_ku^k+u\sum_{k=0}^{n-1}ka_ku^{k-1}=\sum_{k=0}^{n-1}(k-n)a_ku^k,\ a(u)-b(u)=\sum_{k=0}^{n-1}(1+n-k)a_ku^k.$$

By assumption, $a(0)\neq 0$. Thus $a_0\neq 0$, and if we set $\gcd(p,n-k)=d_k$ (where p is the exponent of $\rho(u)=u^p$), then it trivally holds that $d_k=1$ for all k because here we have $p=1$. So $\gcd\left((d_k)_{k|a_k\neq 0}\right)=1$. Moreover

$$\begin{aligned}
\varphi(u)-\frac{u}{\hat{\rho}(\eta)}-\hat{\varphi}(\eta)&=\varphi(u)-u\varphi'(\eta)-(\varphi(\eta)-\eta\varphi'(\eta))\\
&=u^{-n}a(u)-u\eta^{-n-1}b(\eta)-\left(\eta^{-n}a(\eta)-\eta^{-n}b(\eta)\right)\\
&=\frac{1}{u^n\eta^{n+1}}\left(\eta^{n+1}a(u)-u^{n+1}b(\eta)-u^n\eta^{n+1}\left(\eta^{-n}a(\eta)-\eta^{-n}b(\eta)\right)\right)\\
&=\frac{1}{u^n\eta^{n+1}}\left(\eta^{n+1}a(u)-u^{n+1}b(\eta)-u^n\eta\left(a(\eta)-b(\eta)\right)\right).
\end{aligned}$$

As this rational function has a singularity at $(0,0)$, we blow up the ideal (u,η). For this, let us consider the chart with coordinates (v,η) with $e_0:(v,\eta)\mapsto(v\eta,\eta)$. So the previous expression changes to

$$\frac{1}{v^n\eta^n}\left(a(v\eta)-v^{n+1}b(\eta)-v^n\left(a(\eta)-b(\eta)\right)\right)=:\frac{f(v,\eta)}{v^n\eta^n}.$$

This leads to the computation of $\Psi_{\eta\circ e_1}\left(\mathscr{E}^{f(v,\eta)/v^n\eta^n} \otimes e_0^+\left(R[\eta,\eta^{-1}]\right)\right)$. As e_1 acts trivially on η, the module is equal to $\Psi_\eta\left(\mathscr{E}^{f(v,\eta)/v^n\eta^n} \otimes e_0^+\left(R[\eta,\eta^{-1}]\right)\right)$. This module is supported on $\eta = 0$ by definition. Moreover, according to Proposition 2.1, it is supported at most on the set defined by $f(v,0) = 0$. One can also check that it has no component supported at $v = \infty$, by using the same argument in the other chart $e_1 : (v,\eta) \mapsto (v, v\eta)$ of the blowing-up space (see Lemma 3.3.2.6 and Proposition 2.1). The function f can be written as

$$f(v,\eta) = \sum_{k=0}^{n-1} a_k\eta^k\left(v^k + (n-k)v^{n+1} - (1+n-k)v^n\right).$$

We have

$$\frac{\partial}{\partial v} f(v,0) = a_0 n(n+1)v^{n-1}(v-1) = 0 \Leftrightarrow (v = 1 \vee (v = 0 \wedge n \geq 2))$$

because $a_0 \neq 0, n \neq 0$. In addition, we have $f(1,0) = 0$ and $f(0,0) = a_0 \neq 0$. As a consequence, the polynomial $f(v,0) = a_0\left(1 + nv^{n+1} - (n+1)v^n\right)$ has exactly 1 double root, namely in $v = 1$, and the other roots v_i are simple. The branches of $f(v,\eta)$ at the points $(v_i, 0)$ are thus smooth and transversal to $\eta = 0$. Moreover it is easy to see that $(v-1)^2$ divides $f(v,\eta)$. So $f(v,\eta) = \text{unit} \cdot (v-1)^2$.
Thus, in suitable local coordinates (v',η) centered at the points $(v_i,0)$, the exponent $\frac{f(v,\eta)}{v^n\eta^n}$ can be written as $\frac{g(v')}{\eta^n}$, where

$$g(v') = \begin{cases} v' & \text{at } (v_i,0), \\ v'^2 & \text{at } (1,0). \end{cases}$$

Hence the support of $\Psi_\eta\left(\mathscr{E}^{f(v,\eta)/v^n\eta^n} \otimes e_0^+\left(R[\eta,\eta^{-1}]\right)\right)$ is concentrated in one single point. Thus we can apply Lemma 3.2.3 (see [1, Lemma 5.5]) below to $T = e_0^+\left(R[\eta,\eta^{-1}]\right)$: we apply 3.2.3(1) to the germ of f at each $(v_i,0)$ and 3.2.3(2) to the germ of f at $(1,0)$. Then Theorem 3.2.2 is a consequence of [5, Ex. 5.2.1] and [7, Prop. 3.7]. □

3.2.3 Lemma: Let T be a germ of regular meromorphic connections in coordinates (v',η) with pole on $\eta = 0$ at most and let $n \geq 1$. Then

(1) $\Psi_\eta(\mathscr{E}^{v'/\eta^n} \otimes T) = 0$.

(2) $\Psi_\eta(\mathscr{E}^{v'^2/\eta^n} \otimes T)$ is supported at $v' = 0$ and its germ is isomorphic to the germ at $v' = 0$ of $\Psi_\eta(T \otimes L_{\eta,n})$ with $L_{\eta,n} \cong \left(\mathbb{C}[\eta,\eta^{-1}], \mathrm{d} - \frac{n}{2}\frac{\mathrm{d}\eta}{\eta}\right)$.

(3) Let $h(v',\eta) = v'^2\chi(v',\eta) + \eta^l\vartheta(v',\eta)$, where χ, ϑ are local units and $l \in \mathbb{N}$. Let us assume that $n \geq l+1$. Then $\Psi_\eta\left(\mathscr{E}^{h(v',\eta)/\eta^n} \otimes T\right) = 0$. □

3.3. Extension to abitrary meromorphic functions $\psi \in y^{-1}\mathbb{C}[y^{-1}]$

In this section, we want to decompose the formal convolution of two elementary formal meromorphic connections $\mathrm{El}(\mathrm{Id}, \varphi, R), \mathrm{El}(\mathrm{Id}, \psi, S)$ at the origin of Δ, i.e.,

$$\left(\mathscr{H}^0 \pi_+ \gamma_+ \left((\mathscr{E}^\varphi \otimes R) \boxtimes (\mathscr{E}^\psi \otimes S)\right)\right)_0^\wedge$$

with meromorphic functions $\varphi(x) \in x^{-1}\mathbb{C}[x^{-1}]$, $\psi(y) \in y^{-1}\mathbb{C}[y^{-1}]$, and finite dimensional $\mathbb{C}((x))$- resp. $\mathbb{C}((y))$-vector spaces R resp. S with regular connections having singularities at 0 and ∞.
We may write $(\mathscr{H}^0 f_+ (\mathscr{E}^g \otimes T))_0^\wedge = (\mathrm{El}(\mathrm{Id}, \varphi, R) *_0 \mathrm{El}\,(\mathrm{Id}, \psi, S))_0^\wedge$ with $f := \pi \circ \gamma$ and $g := \varphi(x) + \psi(y)$, $T := R \boxtimes S$. Let us start with the diagram

$$U \xrightarrow{(f,g)} \mathbb{C} \times \mathbb{C} \overset{i}{\hookrightarrow} \mathbb{P}^1 \times \mathbb{P}^1 \overset{i_0}{\hookleftarrow} \Delta \times \mathbb{P}^1$$

that we have mentioned in Section 3.2. Let $\mathscr{M} := i_0^*(i_+(f,g)_+ T)^{\mathrm{an}}$. Then Theorem 3.2.1 tells us that at the origin of Δ, the germs $\mathscr{H}^0 f_+(\mathscr{E}^g \otimes T)$ and $\mathscr{H}^0 p_{1+}(\mathscr{E}^{p_2} \otimes \mathscr{H}^0\mathscr{M})$ have the same formal irregular parts. [6, Example 0.4] shows that this formal irregular part can be computed by determining the singular locus of $\mathscr{H}^0 i_+(f,g)_+ T$ in the neighbourhood of $(0, \infty)$.

3.3.1. The singular locus

To prepare the computation, consider the maps $f = \pi \circ \gamma$ and $g = \varphi(x) + \psi(y)$. In order to use [6, Example 0.4], we have to show that f and g are algebraically independent.

3.3.1.1 Lemma: The functions f, g are algebraically independent.

Proof: Let $P(U,V) \in \mathbb{C}[U,V]$ be a polynomial with $P \circ (f(x,y), g(x,y)) = 0$ for all $(x,y) \in U$. Then we get the derivation

$$\nabla \left(P \circ (f(x,y), g(x,y))\right) = \nabla P(U,V)|_{(U,V)=(f,g)} \circ \begin{pmatrix} y & x \\ \varphi'(x) & \psi'(y) \end{pmatrix} = 0.$$

But outside of the curves C_i which fulfill the equation $\det \begin{pmatrix} y & x \\ \varphi'(x) & \psi'(y) \end{pmatrix} = 0$, this matrix is invertible. This gives $(\nabla P)(U,V)|_{(U,V)=(f,g)} = 0$ on the set $U \setminus \bigcup_i C_i$ which is dense in U. Hence $P(U,V) = \mathit{const}$. This is a contradiction. □

Next we use the diagram

$$U \xrightarrow{(f,g)} \mathbb{C} \times \mathbb{C} \xhookrightarrow{i} \mathbb{P}^1 \times \mathbb{P}^1 \xhookleftarrow{i_0} \Delta \times \mathbb{P}^1$$

from above and compute the singular locus of $\mathscr{H}^0 i_+(f,g)_+T$ in the neighbourhood of $(0,\infty)$, where $T = R \boxtimes S$. Let $X := U$ and $Y \subseteq \mathbb{P}^1 \times \mathbb{P}^1$ be an open set which fullfills the inclusion $i \circ (f,g)(X) \subseteq Y$. Remember that the singular locus of a module $\mathcal{N}$ is determined by the characteristic variety $\mathrm{Ch}(\mathcal{N})$ of $\mathcal{N}$. Next set $\iota := i \circ (f,g)$. We use the following diagram [11, p. 76]:

$$\begin{array}{ccccc} T^*Y & \xleftarrow{p} & T^*Y \times_Y X & \xrightarrow{q} & T^*X \\ \downarrow & & \downarrow & & \downarrow \\ Y & \hookleftarrow & X & \xrightarrow{\sim} & X \end{array}$$

T^*X and T^*Y denote the cotangent spaces of X and Y, $p : T^*Y \times_Y X \to T^*Y$ the canonical projection and $q : T^*Y \times_Y X \to T^*X, (\tau, \underline{x}) \mapsto \tau \circ \mathrm{d}\iota_{\underline{x}}$ with $\tau \in T^*_{\iota(\underline{x})}X$ and $\tau \circ \mathrm{d}\iota_{\underline{x}} : T_{\underline{x}}X \to T_{\iota(\underline{x})} \to \mathbb{C}, v \mapsto (\tau \circ \mathrm{d}\iota_{\underline{x}})(v)$.
Note that as a meromorphic connection, T is holonomic. Moreover, as g is defined on X and has poles on $\{xy = 0\}$, we have $X \subseteq \mathbb{G}_m \times \mathbb{G}_m$, hence $\iota : X \to Y$ is proper because then the inverse maps of f and g map compact sets to compact sets (note that Y is an **open** subset of $\mathbb{P}^1 \times \mathbb{P}^1$). So we can apply [12, Remark 2.5.2] which gives $\mathrm{Ch}\,(\mathscr{H}^0\iota_+T) \subseteq p\,(q^{-1}(\mathrm{Ch}(T)))$. Furthermore, due to [13, Th. 6.4.1], even the equality holds. Hence

$$\mathrm{Ch}\left(\mathscr{H}^0\iota_+T\right) = p\left(q^{-1}(\mathrm{Ch}(T))\right).$$

In order to compute $p\,(q^{-1}(\mathrm{Ch}(T)))$, we have to find the exact action of q. For this, consider the two local isomorphisms

$$\theta : \left(Y \times \mathbb{C}^2\right) \times_Y X \to T^*Y \times_Y X, ((\iota(\underline{x}), \underline{\lambda}), \underline{x}) \mapsto \left(\sum_{j=1}^{2} \lambda_j \mathrm{d}y^j_{\iota(\underline{x})}, \underline{x}\right)$$

$$\phi : \mathbb{C}^2 \times X \to T^*X, (\underline{\xi}, \underline{x}) \mapsto \sum_{i=1}^{2} \xi_i \mathrm{d}x^i_{\underline{x}}$$

with

$\underline{x} = (x_1, x_2) := (x, y)$ (the coordinates on X),
$\underline{y} = (y_1, y_2) := (f(x,y), g(x,y))$ (the coordinates on Y) and
$\underline{\xi} = (\xi_1, \xi_2) \in \mathbb{C}^2, \underline{\lambda} = (\lambda_1, \lambda_2) \in \mathbb{C}^2$.

So we can construct the diagram

$$\begin{array}{ccc} (Y \times \mathbb{C}^2) \times_Y X & \longrightarrow & \mathbb{C}^2 \times X \\ \Big\downarrow \theta & & \Big\downarrow \phi \\ T^*Y \times_Y X & \xrightarrow{q} & T^*X \end{array}$$

Let $\mathrm{Gr}(T)$ denote the graded module associated to a good filtration. As T is a regular meromorphic connection,

$$\mathrm{Ch}(T) \cong \{\xi_1 = \xi_2 = 0\} \cup \{x_1 = 0\} \cup \{x_2 = 0\} \subset \mathbb{C}^2 \times X = \phi^{-1}(T^*X),$$

where ξ_1, ξ_2 are the symbols associated to $\partial_{x_1}, \partial_{x_2}$ in $\mathrm{Gr}(T)$. So we have

$$\phi^{-1}(\mathrm{Ch}(T)) = \{\xi_1 = \xi_2 = 0\} \cup \{x_1 = 0\} \cup \{x_2 = 0\}.$$

Next we want to compute $(\theta^{-1} \circ q^{-1} \circ \phi)\,(\phi^{-1}(\mathrm{Ch}(T))) \subseteq (Y \times \mathbb{C}^2) \times_Y X$.
For $((\iota(\underline{x}), \underline{\lambda}), \underline{x}) \in (Y \times \mathbb{C}^2) \times_Y X$, it is

$$\theta((\iota(\underline{x}), \underline{\lambda}), \underline{x}) = \left(\sum_{j=1}^{2} \lambda_j \mathrm{d}y^j_{\iota(\underline{x})}, \underline{x}\right) \in T^*_{\iota(\underline{x})}Y \times_Y \{\underline{x}\}.$$

Then, for $v \in T_{\underline{x}}X$,

$$q\left(\sum_{j=1}^{2} \lambda_j \mathrm{d}y^j_{\iota(\underline{x})}, \underline{x}\right)(v) \overset{(*)}{=} \underbrace{\left(\sum_{j=1}^{2} \lambda_j \mathrm{d}y^j_{\iota(\underline{x})}\right)}_{\in T^*_{\iota(\underline{x})}Y} \underbrace{(\mathrm{d}\iota_{\underline{x}}(v))}_{\in T_{\iota(\underline{x})}Y}.$$

On the other hand,

$$q\left(\sum_{j=1}^{2} \lambda_j \mathrm{d}y^j_{\iota(\underline{x})}, \underline{x}\right) = \sum_{i=1}^{2} \xi_i \mathrm{d}x^i_{\underline{x}} \in T^*_{\underline{x}}X.$$

Now we want to express the "variables" ξ_i in terms of the λ_j.
We know that $\left\{ \frac{\partial}{\partial x_1}\Big|_{\underline{x}}, \frac{\partial}{\partial x_2}\Big|_{\underline{x}} \right\} \subset T_{\underline{x}}X$ is a dual basis of $\left\{\mathrm{d}x^1_{\underline{x}}, \mathrm{d}x^2_{\underline{x}}\right\} \subset T^*_{\underline{x}}X$ and that there is an isomorphism $\mathbb{C}^2 \cong T_{\underline{x}}X (\ni v)$ via the mapping $(\xi_1, \xi_2) \mapsto \sum_{i=1}^{2} \xi_i \frac{\partial}{\partial x_i}\Big|_{\underline{x}}$.

So, with $v_k := \left.\frac{\partial}{\partial x_k}\right|_{\underline{x}}$, this gives

$$\xi_k = \left(\sum_{i=1}^{2} \xi_i \mathrm{d}x^i_{\underline{x}}\right)(v_k) = q\left(\sum_{j=1}^{2} \lambda_j \mathrm{d}y^j_{\iota(\underline{x})}, \underline{x}\right)(v_k) \overset{(*)}{=} \left(\sum_{j=1}^{2} \lambda_j \mathrm{d}y^j_{\iota(\underline{x})}\right)(\mathrm{d}\iota_{\underline{x}}(v_k)).$$

Hence, expressed in terms of the basis $\left\{ \left.\frac{\partial}{\partial_{y_1}}\right|_{\iota(\underline{x})}, \left.\frac{\partial}{\partial_{y_2}}\right|_{\iota(\underline{x})} \right\}$ of $T_{\iota(\underline{x})}Y$,

$$\xi_1 = \left(\sum_{j=1}^{2} \lambda_j \mathrm{d}y^j_{\iota(\underline{x})}\right)\left(\frac{\partial f}{\partial x_1}\left.\frac{\partial}{\partial y_1}\right|_{\iota(\underline{x})} + \frac{\partial g}{\partial x_1}\left.\frac{\partial}{\partial y_2}\right|_{\iota(\underline{x})}\right) = \frac{\partial f}{\partial x_1}\lambda_1 + \frac{\partial g}{\partial x_1}\lambda_2$$
$$\xi_2 = \left(\sum_{j=1}^{2} \lambda_j \mathrm{d}y^j_{\iota(\underline{x})}\right)\left(\frac{\partial f}{\partial x_2}\left.\frac{\partial}{\partial y_1}\right|_{\iota(\underline{x})} + \frac{\partial g}{\partial x_2}\left.\frac{\partial}{\partial y_2}\right|_{\iota(\underline{x})}\right) = \frac{\partial f}{\partial x_2}\lambda_1 + \frac{\partial g}{\partial x_2}\lambda_2$$

This allows us to define the map $\phi^{-1} \circ q \circ \theta$ and to determine $\mathrm{Ch}(\mathscr{H}^0\iota_+T)$:

$$\begin{aligned}\phi^{-1} \circ q \circ \theta : \left(Y \times \mathbb{C}^2\right) \times_Y X &\to \mathbb{C}^2 \times X \\ \left(\left(\iota(\underline{x}), \underline{\lambda}\right), \underline{x}\right) &\mapsto \left(\begin{pmatrix} \frac{\partial f}{\partial x_1}\lambda_1 + \frac{\partial g}{\partial x_1}\lambda_2 \\ \frac{\partial f}{\partial x_2}\lambda_1 + \frac{\partial g}{\partial x_2}\lambda_2 \end{pmatrix}, \underline{x}\right)\end{aligned}$$

As $\phi^{-1}(\mathrm{Ch}(T)) = \{\xi_1 = \xi_2 = 0\} \cup \{x_1 = 0\} \cup \{x_2 = 0\}$,

$$\begin{aligned}&\left(\theta^{-1} \circ q^{-1} \circ \phi\right)\left(\phi^{-1}(\mathrm{Ch}(T))\right) \\ &= \left\{ \left(\theta^{-1} \circ q^{-1} \circ \phi\right)\left(\begin{pmatrix} 0 \\ 0 \end{pmatrix}, \underline{x}\right) \middle| \underline{x} \in X \right\} \\ &\cup \left\{ \left(\theta^{-1} \circ q^{-1} \circ \phi\right)\left(\underline{\xi}, \begin{pmatrix} x_1 \\ x_2 \end{pmatrix}\right) \middle| x_1 = 0 \vee x_2 = 0 \right\} \\ &= \left\{ \left(\left(\iota\left(\underline{x}\right), \underline{\lambda}\right), \underline{x}\right) \middle| \begin{pmatrix} \frac{\partial f}{\partial x_1} & \frac{\partial g}{\partial x_1} \\ \frac{\partial f}{\partial x_2} & \frac{\partial g}{\partial x_2} \end{pmatrix} \underline{\lambda}^T = \begin{pmatrix} 0 \\ 0 \end{pmatrix}, \underline{x} \in X \right\} \\ &\cup \left\{ \left(\left(\iota\left(\underline{x}\right), \underline{\lambda}\right), \underline{x}\right) \middle| \begin{pmatrix} \frac{\partial f}{\partial x_1} & \frac{\partial g}{\partial x_1} \\ \frac{\partial f}{\partial x_2} & \frac{\partial g}{\partial x_2} \end{pmatrix} \underline{\lambda}^T \in \mathbb{C}^2, \underline{x} \in X, x_1 = 0 \vee x_2 = 0 \right\}.\end{aligned}$$

Next we extend the diagram of page 22:

$$\begin{array}{ccccc} Y \times \mathbb{C}^2 & \xleftarrow{\tilde{p}} & (Y \times \mathbb{C}^2) \times_Y X & \longrightarrow & \mathbb{C}^2 \times X \\ \downarrow \tilde{\theta} & & \downarrow \theta & & \downarrow \phi \\ T^*Y & \xleftarrow{p} & T^*Y \times_Y X & \xrightarrow{q} & T^*X \end{array}$$

with the canonical projection $\tilde{p}$ and the isomorphism $\tilde{\theta}$ induced by θ. This gives

$$\mathrm{Ch}\left(\mathscr{H}^0 \iota_+ T\right) = \tilde{\theta} \circ \tilde{p} \circ \theta^{-1} \circ q^{-1}\left(\mathrm{Ch}(T)\right)$$

and above computations showed that, with the help of the isomorphism $\tilde{\theta}$, this set can also be represented as

$$\begin{aligned} \tilde{\theta}^{-1}\mathrm{Ch}\left(\mathscr{H}^0 \iota_+ T\right) = & \left\{ \left(\iota\left(\underline{x}\right), \underline{\lambda}\right) \middle| \begin{pmatrix} \frac{\partial f}{\partial x_1} & \frac{\partial g}{\partial x_1} \\ \frac{\partial f}{\partial x_2} & \frac{\partial g}{\partial x_2} \end{pmatrix} \underline{\lambda}^T = \begin{pmatrix} 0 \\ 0 \end{pmatrix}, \underline{x} \in X \right\} \\ & \cup \left\{ \left(\iota\left(\underline{x}\right), \underline{\lambda}\right) \middle| \begin{pmatrix} \frac{\partial f}{\partial x_1} & \frac{\partial g}{\partial x_1} \\ \frac{\partial f}{\partial x_2} & \frac{\partial g}{\partial x_2} \end{pmatrix} \underline{\lambda}^T \in \mathbb{C}^2, \underline{x} \in X, x_1 = 0 \vee x_2 = 0 \right\} \end{aligned}$$

as a subset of $Y \times \mathbb{C}^2$. But the introduction of [6] tells us just to consider the components of the characteristic variety distinct of $\{x_1 = 0 \vee x_2 = 0\}$. Hence we just have to consider the zero section

$$B := \left\{ \left(\iota\left(\underline{x}\right), \underline{\lambda}\right) \middle| \begin{pmatrix} \frac{\partial f}{\partial x_1} & \frac{\partial g}{\partial x_1} \\ \frac{\partial f}{\partial x_2} & \frac{\partial g}{\partial x_2} \end{pmatrix} \underline{\lambda}^T = \begin{pmatrix} 0 \\ 0 \end{pmatrix}, \underline{x} \in X \right\} \subseteq \tilde{\theta}^{-1}\mathrm{Ch}\left(\mathscr{H}^0 \iota_+ T\right).$$

Let us define the matrix

$$A := \begin{pmatrix} \frac{\partial f}{\partial x_1} & \frac{\partial g}{\partial x_1} \\ \frac{\partial f}{\partial x_2} & \frac{\partial g}{\partial x_2} \end{pmatrix}.$$

With the help of A, we are able to compute the **singular support** of $\mathscr{H}^0 \iota_+ T$. This can be done as follows: Let $\underline{y} = (y_1, y_2) = i(f(x_1, x_2), g(x_1, x_2)) \in Y$. Then a module is defined to be singular at $\underline{y}$ if

$$\mathrm{Ch}\left(\mathscr{H}^0 i_+(f,g)_+ T\right) \cap T^*_{\underline{y}} Y \neq \{0\}.$$

But $\mathrm{Ch}\left(\mathscr{H}^0 i_+(f,g)_+ T\right)$ is contained in T^*Y. So the *singular support* is defined by

$$\begin{aligned} \mathrm{Supp}\ \left(\mathscr{H}^0 i_+(f,g)_+ T\right) := & \left\{ \underline{y} \in Y | \mathrm{Ch}\left(\mathscr{H}^0 i_+(f,g)_+ T\right) \cap T^*_{\underline{y}} Y \neq \{0\} \right\} \\ = & \left\{ \underline{y} \in Y | \mathrm{Ch}\left(\mathscr{H}^0 i_+(f,g)_+ T\right)_{\underline{y}} \neq \{0\} \right\} \end{aligned}$$

(see [10, (2.4.3)]). Then, as we are just interested in the subset $B \subseteq \tilde{\theta}^{-1}\mathrm{Ch}(\mathscr{H}^0\iota_+T)$,

$$\begin{aligned} C &:= \left\{(x_1, x_2) \in X | i(f,g)(x_1,x_2) \in \mathrm{Supp}\left(\mathscr{H}^0 i_+(f,g)_+T\right) \cap B\right\} \\ &= \left\{(x_1,x_2) \in X | \det(A) = 0\right\}. \end{aligned}$$

Now this fact leads directly to the computation of the **singular locus** of the module $\mathscr{H}^0 i_+(f,g)_+T$. For this, we use [6, Example 0.4] again: In a neighbourhood of the point $(0,\infty) \in \Delta \times \mathbb{P}^1$, when we denote the singular locus by $\mathscr{S}$, $\mathscr{S}$ consists of the set $\{\mathscr{S}_k | k \in \Lambda\}$ of local irreducible components which are distinct of $\{0\} \times \mathbb{P}^1$ or $\Delta \times \{\infty\}$. As we consider the multiplicative convolution

$$\mathscr{H}^0 \pi_+ \gamma_+ \left(\mathscr{E}^{\varphi(x)+\psi(y)} \otimes T\right) = \mathscr{H}^0 f_+ \left(\mathscr{E}^g \otimes T\right),$$

we have

$$f = \pi \circ \gamma : (x,y) \mapsto xy, \text{ and } g = \varphi + \psi : (x,y) \mapsto \varphi(x) + \psi(y).$$

So

$$C = \left\{(x,y) \middle| \det \begin{pmatrix} \frac{\partial f(x,y)}{\partial x} & \frac{\partial g(x,y)}{\partial x} \\ \frac{\partial f(x,y)}{\partial y} & \frac{\partial g(x,y)}{\partial y} \end{pmatrix} = 0 \right\} = \left\{(x,y) \left| x\varphi'(x) - y\psi'(y) = 0\right.\right\}.$$

Then finally, the components $\mathscr{S}_k$ of the singular locus $\mathscr{S}$ are defined to be the irreducible components at $(0,\infty)$ of the germ

$$\overline{(f,g)(C)} \setminus \left(\left(\{0\} \times \mathbb{P}^1\right) \cup \left(\Delta \times \{\infty\}\right)\right).$$

Thus we have proved the following:

3.3.1.2 Proposition: Let $f(x,y)$ and $g(x,y)$ be algebraically independent regular functions. Then the singular locus of the module $\mathscr{H}^0 i_+(f,g)_+T$ is the set

$$\begin{aligned} &\mathrm{Supp}\left(\mathscr{H}^0 i_+(f,g)_+T\right) \setminus \left(\left(\{0\} \times \mathbb{P}^1\right) \cup \left(\Delta \times \{\infty\}\right)\right) \\ &= \overline{(f,g)(C)} \setminus \left(\left(\{0\} \times \mathbb{P}^1\right) \cup \left(\Delta \times \{\infty\}\right)\right) \end{aligned}$$

with

$$C = \left\{(x,y) \middle| \det \begin{pmatrix} \frac{\partial f(x,y)}{\partial x} & \frac{\partial g(x,y)}{\partial x} \\ \frac{\partial f(x,y)}{\partial y} & \frac{\partial g(x,y)}{\partial y} \end{pmatrix} = 0 \right\}$$

□

3.3.2. The convolution formula

Now we are ready to compute the formalization of $\mathrm{El}(\mathrm{Id}, \varphi, R) *_0 \mathrm{El}(\mathrm{Id}, \psi, S)$ at the origin of Δ. According to Proposition 3.3.1.2, $C = \{(x, y) | y\psi'(y) = x\varphi'(x)\}$. We define

$$\varphi(x) := \sum_{i=1}^{n} a_i \frac{1}{x^i} \quad \text{and} \quad \psi(y) := \sum_{j=1}^{m} b_j \frac{1}{y^j}$$

where $a_n \neq 0$ and $b_m \neq 0$. This means, we have to consider the set of pairs (x, y) with

$$x\varphi'(x) - y\psi'(y) = \sum_{i=1}^{n} -ia_i \frac{1}{x^i} + \sum_{j=1}^{m} jb_j \frac{1}{y^j} = 0.$$

Applying γ_+, where $\gamma : (x, y) \mapsto (x, xy) =: (u, z)$, this gives the constraint

$$\sum_{i=1}^{n} -ia_i \frac{1}{u^i} + \sum_{j=1}^{m} jb_j \frac{u^j}{z^j} = 0.$$

Next we want to find parametrizations which fulfill this equation - the so called *Puiseux parametrizations.* This means, we want to find parametrizations of the set

$$C' := \left\{ (u, z) \middle| \sum_{i=1}^{n} -ia_i \frac{1}{u^i} + \sum_{j=1}^{m} jb_j \frac{u^j}{z^j} = 0 \right\}.$$

Remember that we have to express z in terms of u (and not u in terms of z) because the convolution contains the projection $\pi : (u, z) \mapsto z$.

3.3.2.1 Lemma: The set $\left\{ (u, z) \middle| \sum_{i=1}^{n} -ia_i \frac{1}{u^i} + \sum_{j=1}^{m} jb_j \frac{u^j}{z^j} = 0 \right\}$ is parameterized by exactly m pairs of the form

$$(u, z_k(u)) = \left(u, \sum_{p=1}^{\infty} c_{p,k} u^{\frac{n+m+\gcd(m,n)(p-1)}{m}} \right), k = 1, ..., m.$$

Proof: First we multiply the equation

$$\sum_{i=1}^{n} -ia_i \frac{1}{u^i} + \sum_{j=1}^{m} jb_j \frac{u^j}{z^j} = 0$$

by $u^n z^m$. Then it is equivalent to the equation $P(u,z) = 0$ with

$$P(u,z) = mb_m u^{n+m} + (m-1)b_{m-1}u^{n+m-1}z + ... + b_1 u^{n+1} z^{m-1} + \left(\sum_{i=1}^{n} -ia_i u^{n-i} \right) z^m.$$

As we want to find parametrizations $(u, z(u))$, we consider $P(u,z)$ as a polynomial in z with coefficients in $\mathbb{C}[u]$. For this purpose, we use the construction in [14, Sect. 2.5] (note that in the all computations of this work, slopes are strictly negative; nevertheless we will always use their absolute values):

The first step is to consider the coefficients of z^j $(j \in \{0, ..., m\})$. Pick out the lowest u-exponent s_j of every coefficient, and draw the points (j, s_j) in a coordinate system (if one coefficient is equal to zero, then there is no point to mark). One can easily see that the connection of the points in this first step never gives a convex graph. Thus the convex hull (the Newton polygon) of the points (j, s_j) will always have the single slope $\frac{m+n}{m}$ different from 0 and ∞ (no matter how many of the coefficients besides the ones of z^0 and z^m are equal to zero). Hence $z(u) = c_1 u^{(m+n)/m} + \tilde{z}_1$ with a new variable $\tilde{z}_1$. Next we consider $P\left(u, c_1 u^{(m+n)/m} + \tilde{z}_1\right)$ and choose c_1 such that the term with the lowest u-exponent in the coefficient of $\tilde{z}_1^0$ vanishes. This gives $c_{1,k} = \zeta_m^k \sqrt[m]{\frac{mb_m}{na_n}}$, $k \in \{1, ..., m\}$ (note that $a_n, b_m \neq 0$). So we have found m starting terms which lead to parametrizations $z_k(u)$ for $k = 1, ..., m$ due to [14]. As $P(u,z)$ is a polynomial of degree m in z, there can be no more parametrizations.

Now we want to show that during the further procedure all Newton polygons have exactly two slopes - the used slope $\frac{m+n}{m}$ and a new slope strictly greater than $\frac{m+n}{m}$ which is an integer multiple of $\frac{1}{m^*}$ with $m^* := \frac{m}{\gcd(m,n)}$. We will prove this by induction over the number q of executed steps of the algorithm. For the sake of simplicity, we leave out the index $k \in \{1, ..., m\}$ (because we already know that there are exactly m parametrizations) and just use the index q. Before starting the induction, we need some notation: We denote by $P_1(u, \tilde{z}_1) := P\left(u, \zeta_m \sqrt[m]{\frac{mb_m}{na_n}} u^{(m+n)/m} + \tilde{z}_1\right) \in \left(\mathbb{C}[u^{1/m}]\right)[\tilde{z}_1]$ the function which we receive after the construction of the first Newton polygon at the beginning of the proof. In analogy, we denote by $P_q(u, \tilde{z}_q) := P_{q-1}\left(u, c_q u^{\gamma_q} + \tilde{z}_q\right)$ the function which we receive after the construction of the q-th Newton polygon.

We want to show that for all $q \in \mathbb{N}$, the function $P_q(u, \tilde{z}_q)$ has exactly the two slopes $\frac{m+n}{m}$ and $\gamma_{q+1} > \gamma_q$, where γ_{q+1} is an integer multiple of $\frac{1}{m^*}$. To do this, we show the following three points:

1) For all $q \in \mathbb{N}$, the lowest u-exponents in the coefficients of $\tilde{z}_q^j$ in the function $P_q(u, \tilde{z}_q)$ are $m + n - j\frac{m+n}{m}$, $j \in \{1, ..., m\}$.

2) For all $q \in \mathbb{N}$, all u-exponents in the coefficient of $\tilde{z}_q^0$ in the function $P_q(u, \tilde{z}_q)$ are integer multiples of $\frac{1}{m^*}$ greater than $\gamma_q + m + n - \frac{m+n}{m}$ (then we have $\gamma_{q+1} > \gamma_q$).

3) For all $q \in \mathbb{N}$, the defining equation for the coefficient c_{q+1} is linear in c_{q+1}, so c_{q+1} is uniquely determined.

For the induction basis, consider $P_1(u, \tilde{z}_1) = P\left(u, \zeta_m \sqrt[m]{\frac{mb_m}{na_n}} u^{(m+n)/m} + \tilde{z}_1\right)$. Computing $P_1(u, \tilde{z}_1)$, we get a function in which the coefficient of $\tilde{z}_1^j$ for $j = 0, ..., m-1$ consists of a sum of terms with u-exponents $m + n - j - l + l\frac{m+n}{m}$ and $n - i + (m-j)\frac{m+n}{m}$ for $l \in \{0, ...m - j - 1\}$, $i \in \{1, ..., n\}$. Note that, due to the construction, for $j = 0$ the terms with u-exponent $m + n$ vanish, so the lowest u-exponent in the coefficient of $\tilde{z}_1^0$ is t_1 which is an integer of $\frac{1}{m^*}$ and greater than $m + n$. Moreover, as $P(u, z)$ only contained integer u-exponents, condition 2) is clearly true. The lowest u-exponent in the coefficient of $\tilde{z}_1^m$ can directly seen to be 0. It remains to find the lowest u-exponent in every coefficient of the $\tilde{z}_1^j$ for $j = 1, ..., m-1$. They are given by $m + n - j(\frac{m+n}{m})$ - these are exactly the u-terms which contain $a_n \neq 0$ in the coefficient (besides other non-zero factors), hence they are not zero. Thus also condition 1) holds. So the Newton polygon of $P_1(u, \tilde{z}_1)$ has the two slopes $\frac{m+n}{m}$ and $\gamma_2 := t_1 - \left(m + n - \frac{m+n}{m}\right) > \frac{m+n}{m} = \gamma_1$ (as $t_1 > m + n$). Now let

$$\bar{P}_1(u, \tilde{z}_1) := u^{t_1} + u^{m+n-\frac{m+n}{m}} \tilde{z}_1 + ... + u^{m+n-(m-1)\frac{m+n}{m}} \tilde{z}_1^{m-1} + \tilde{z}_1^m$$

be the auxiliary function which contains only the terms that define the Newton polygon and no constants. Set $\tilde{z}_1 := c_2 u^{\gamma_2} + \tilde{z}_2$. The function $\bar{P}_2(u, \tilde{z}_2) := \bar{P}_1(u, c_2 u^{\gamma_2} + \tilde{z}_2)$ suffices to compute c_2 because c_2 only has to fulfill the property that the term with the lowest u-exponent t_1 in the coefficient of $\tilde{z}_2^0$ in $P_2(u, \tilde{z}_2)$ vanishes. In $\bar{P}_2(u, \tilde{z}_2)$, the coefficient of $\tilde{z}_2^0$ consists of terms with the u-exponents t_1 and $m + n - j\frac{m+n}{m} + j\gamma_2$ for $j = 1, ..., m$. But

$$m + n - j\frac{m+n}{m} + j\gamma_2 = m + n + j(t_1 - m - n) > t_1 \text{ if } j > 1.$$

This means, the defining equation for c_2 is linear in c_2, i.e., an equation of the form $a + c_2 b = 0$ for $a, b \in \mathbb{C}$. Hence c_2 is uniquely determined. This proves condition 3).
For the induction step, we want to show that the function $P_q(u, \tilde{z}_q)$ fulfills the three conditions. By definition, $P_q(u, \tilde{z}_q) = P_{q-1}(u, c_q u^{\gamma_q} + \tilde{z}_q)$. By the induction hypothesis, $P_{q-1}(u, \tilde{z}_{q-1})$ has the three desired properties. As $P_q(u, \tilde{z}_q)$ is given, we know $c_q u^{\gamma_q}$. So

$$\begin{aligned} P_q(u, \tilde{z}_q) =& P_{q-1}(u, c_q u^{\gamma_q} + \tilde{z}_q) = f_{q-1}\left(u^{\frac{1}{m^*}}\right) + \left(d_1 u^{m+n-\frac{m+n}{m}} + ...\right)(c_q u^{\gamma_q} + \tilde{z}_q) + ... \\ &+ \left(d_{m-1} u^{m+n-(m-1)\frac{m+n}{m}} + ...\right)(c_q u^{\gamma_q} + \tilde{z}_q)^{m-1} - \left(\sum_{i=1}^{n} i a_i u^{n-i}\right)(c_q u^{\gamma_q} + \tilde{z}_q)^m \end{aligned}$$

with a polynomial $f_{q-1}(s) \in \mathbb{C}[s]$ such that $f_{q-1}\left(u^{\frac{1}{m^*}}\right)$ has property 2) and $d_j \in \mathbb{C}^\times$ for all $j \in \{1, ..., m-1\}$. Considering the coefficients of $\tilde{z}_q^j$ for $j \in \{1, ..., m\}$, the lowest u-exponents are $m+n-j\frac{m+n}{m}$ again, thus $P_q(u, \tilde{z}_q)$ has property 1). The coefficient of $\tilde{z}_q^0$ contains $f\left(u^{\frac{1}{m^*}}\right)$ (which has u-exponents that are greater than $\gamma_{q-1}+m+n-\frac{m+n}{m}$ due to property 1)) and new terms with u-exponents $m+n-j\frac{m+n}{m}+j\gamma_q > \gamma_{q-1}+m+n-\frac{m+n}{m}$ (and greater ones), hence $P_q(u, \tilde{z}_q)$ has property 2) because by the induction hypothesis, γ_q is an integer multiple of $\frac{1}{m^*}$ and because the term with u-exponent $\gamma_{q-1}+m+n-\frac{m+n}{m}$ vanishes due to the choice of c_q. This gives the fact that the Newton polygon has the two slopes $\frac{m+n}{m}$ and $\gamma_{q+1} > \gamma_q$ with $\gamma_{q+1} \in \frac{1}{m^*}\mathbb{N}$. Finally property 3) holds which can be shown as in the induction basis. By induction, we get the desired parametrizations

$$(u, z_k(u)) = \left(u, \sum_{p=1}^{\infty} c_{p,k} u^{\frac{m+n+\gcd(n,m)(p-1)}{m}}\right), \quad k = 1, ..., m.$$

□

Now let $d := \gcd(n, m)$ and consider $\frac{m}{d} := m^*, \frac{n}{d} := n^*$. We will use the function $\beta : t \mapsto t^{m^*} = u$ in order to equip the $z_k(u)$ with integer exponents. Hence C' is parameterized by the set

$$C'' := \left\{\left(t^{m^*}, z_k\left(t^{m^*}\right)\right) \middle| \, k \in \{1, ..., m\}\right\}.$$

To complete the parametrization of the singular locus, we need the **irreducible** components of the germ

$$\overline{\left(\tilde{f}, \tilde{g}\right)(C'')} \setminus \left(\left(\{0\} \times \mathbb{P}^1\right) \cup (\Delta \times \{\infty\})\right)$$

at $(0, \infty)$ with $\tilde{f}(u, z) := z, \tilde{g}(u, z) := \varphi(u) + \psi\left(\frac{z}{u}\right)$. First of all, this means that we need a decomposition $P(u, z) = \prod_{i=1}^r P_i^{s_i}$ of the polynomial $P(u, z)$ (see the proof of Lemma 3.3.2.1) into pairwise distinct irreducible components $P_1, ..., P_r$ over $\mathbb{C}[\![u]\!][z]$. Then for every P_i, we choose one corresponding parametrization which is an element of the set C'' from above due to Lemma 3.3.2.1. Hence the singular locus is the set

$$\left\{\left(z_k\left(t^{m^*}\right), \varphi\left(t^{m^*}\right) + \psi\left(\frac{z_k\left(t^{m^*}\right)}{t^{m^*}}\right)\right) \middle| \, k \in \{1, ..., r\}\right\}.$$

3.3.2.2 Remark: A decomposition of $P(u, z)$ in irreducible components over $\mathbb{C}[\![u]\!][z]$ cannot be found that easily in general. However there are some interesting cases:

1) In the case $m = n$, Lemma 3.3.2.1 shows that all m parametrizations of

C' have integer exponents. Hence $P(u,z)$ can be decomposed into $r = m$ irreducible components over $\mathbb{C}[\![u]\!][z]$.

2) In the case $\gcd(m,n) = 1$, $P(u,z)$ is irreducible over $\mathbb{C}[\![u]\!][z]$, i.e., $r = 1$ (see [15, Corollary 3.2]).

3) In order to find a decomposition of P in the general case, one can use [16].

Next we want to bring the parametrizations in another form in order to apply the main result of [6]. Let $\tilde{\rho}_k(t) := z_k\left(t^{m^*}\right)$. Now choose for every $k \in \{1,...,r\}$ a power series $\lambda_k(t) \in t\mathbb{C}[\![t]\!]$ with $\lambda_k'(0) \neq 0$ and $(\tilde{\rho}_k \circ \lambda_k)(t) = t^{m^*+n^*}$. Such a series can be constructed step by step. We will see later that only finitely many summands of the z_k and the λ_k have to be computed. Set $\rho(t) := (\tilde{\rho}_k \circ \lambda_k)(t) = t^{m^*+n^*}$. The same transformation has to be applied to the other component of the parametrizations, i.e., we have to consider

$$\varphi\left((\lambda_k(t))^{m^*}\right) + \psi\left(\frac{t^{m^*+n^*}}{(\lambda_k(t))^{m^*}}\right) = \sum_{i=1}^{n} a_i \frac{1}{(\lambda_k(t))^{m^*i}} + \sum_{j=1}^{m} b_j \frac{(\lambda_k(t))^{m^*j}}{t^{j(m^*+n^*)}}, \quad k \in \{1,...,r\}.$$

By polynomial division one can show that these functions are equal to the Laurent series

$$\underbrace{\sum_{l=1}^{m^*n} d_{l,k} t^{-l}}_{=:\alpha_k(t)\text{ (meromorphic part)}} + \underbrace{\sum_{l=0}^{\infty} \tilde{d}_{l,k} t^{l}}_{=:\delta_k(t)\text{ (holomorphic part)}}, \quad k \in \{1,...,r\}.$$

Now it still can happen that the set

$$\{(\rho(t), \alpha_k(t) + \delta_k(t)),\ k = 1,...,r\}$$

contains too many parametrizations, i.e., that the representation of this set is not minimal. This is the case if there is a $(m^* + n^*)$-th root of unity ζ such that we have an equality of the form $\alpha_k(\zeta t) + \delta_k(\zeta t) = \alpha_l(t) + \delta_l(t)$ for $k,l \in \{1,...,r\}$ with $k \neq l$. If we cancel these "redundant" parametrizations, we get (after renumbering the parametrizations, if necessary) the minimal representation

$$\mathscr{S} := \{(\rho(t), \alpha_k(t) + \delta_k(t)),\ k = 1,...,r^*\}.$$

of the singular locus. Clearly $r^* \leq r$.

3.3.2.3 Remark: In the case $\gcd(n,m) = 1$, we also have $r^* = r = 1$.

As a consequence, the parametrizations are in the right form for the following statement. It is the main statement of [6] and gives a first decomposition formula.

3.3.2.4 Theorem [6, Th. 0.1]: Let

(i) T be a regular holonomic $\mathscr{D}_{\Delta\times\mathbb{P}^1}$-module with a disc Δ centered at the origin (remember that in our case, we have $T = R \boxtimes S$).

(ii) $p_1 : \Delta \times \mathbb{P}^1 \to \Delta$ and $p_2 : \Delta \times \mathbb{P}^1 \to \mathbb{P}^1$ (with polar divisor $\Delta \times \{\infty\}$) be the canonical projections.

(iii) $\mathscr{S}$ be the singular locus of T in a neighbourhood U of $(0,\infty)$. Moreover let $\{\mathscr{S}_k | k \in \{1,...,r^*\}\}$ be the set consisting of the local irreducible components of $\mathscr{S}$ which are distinct of $\{0\} \times \mathbb{P}^1$ or $\Delta \times \{\infty\}$.

(iv) $\alpha_k \in t^{-1}\mathbb{C}[t^{-1}]$ and $\delta_k \in \mathbb{C}\{t\}$ the polar part and the holomorphic part of a Puiseux parametrization of $\mathscr{S}_k$ at $(0,\infty)$ such that $(t^{p_k}, \alpha_k(t) + \delta_k(t))$ is a parametrization of $\mathscr{S}_k$ (in our case, we have $p_1 = p_2 = ... = p_m = m^* + n^*$; also the pole orders of the α_k are identical for all $k \in \{1,...,r^*\}$).

Using the ramification $\rho(t) = t^p$ with $p = \operatorname{lcm}(p_1,...,p_m)$ (here $p = m^* + n^*$), the formalization of $\rho^+\mathscr{H}^0 p_{1+}(\mathscr{E}^{p_2} \otimes T)_0$ decomposes as $\bigoplus_{\alpha\in\Gamma} \mathscr{E}^{\alpha} \otimes T_\alpha$, where $\Gamma \subset t^{-1}\mathbb{C}[t^{-1}]$ is a finite set and $\alpha \in \Gamma$ if and only if there exists $k \in \{1,...,r^*\}$ and $\zeta \in \mathbb{C}^\times$ with $\zeta^p = 1$ such that $\alpha(t) = \alpha_k(\zeta t^{p/p_k})$; the set of such $k \in \{1,...,r^*\}$ is denoted by Λ_α.
T_α is regular holonomic with $T_\alpha = T_\alpha[t^{-1}] \neq 0$ and $\operatorname{rk}(T_\alpha) = \sum_{k\in\Lambda_\alpha} m_k$, where $m_k \in \mathbb{N}$ denotes the multiplicity of the conormal space $T^*_{\mathscr{S}_k}X$ in the characteristic cycle of T. □

Note that this theorem does not distinguish between functions $\alpha(t)+\delta(t)$ and $\alpha(t)+\tilde{\delta}(t)$ with $\delta(t) \neq \tilde{\delta}(t)$. This may lead to missunderstandings in Theorem 3.3.2.4 and is the main difficulty in the computation of the convolution.
Considering the pairs $(\rho(t), \alpha_k(t)), k \in \{1,...,r^*\}$ and Theorem 3.2.1, we can apply Theorem 3.3.2.4. The result is a decomposition

$$\left(\rho^+ \left(\mathrm{El}(\mathrm{Id}, \varphi, R) *_0 \mathrm{El}\left(\mathrm{Id}, \psi, S\right)\right)\right)_0^{\wedge,\mathrm{loc.}} \cong \bigoplus_{\alpha\in\Gamma} \mathscr{E}^{\alpha} \otimes T_\alpha$$

with

$$\Gamma := \left\{\alpha_k(\xi t) \,|\xi \in \mathbb{C}^* : \xi^{m^*+n^*} = 1; k = 1,...,r^*\right\}.$$

Γ contains at most $r^*(m^* + n^*) \leq m(m^* + n^*)$ distinct elements. Note that, as we only need the meromorphic parts α_k, we only need to determine finitely many summands of the z_k and λ_k. So, until now, we have shown a first decomposition formula for the local formal convolution. We will summarize these results in the following theorem.

3.3.2.5 Theorem: Let C' be the set defined on page 26 and $(u, z_k(u))$ with $k = 1, ..., r$ be the corresponding parametrizations of the pairwise distinct irreducible components of a decomposition of the polynomial $P(u, z)$ over $\mathbb{C}[\![u]\!][z]$ which defines C' (see Lemma 3.3.2.1). Then the set

$$\left\{\left(z_k\left(t^{m^*}\right), \varphi\left(t^{m^*}\right) + \psi\left(\frac{z_k\left(t^{m^*}\right)}{t^{m^*}}\right)\right)\middle| k \in \{1, ..., r\}\right\}$$

is the singular locus. After setting $\tilde{\rho}_k(t) := z_k\left(t^{m^*}\right)$ and choosing $\lambda_k(t) \in t\mathbb{C}[\![t]\!]$ with $\lambda'(0) \neq 0$ and $\rho(t) := (\tilde{\rho}_k \circ \lambda_k)(t) = t^{m^*+n^*}$, the set

$$\left\{\left(\rho(t), \varphi\left((\lambda_k(t))^{m^*}\right) + \psi\left(\frac{t^{m^*+n^*}}{(\lambda_k(t))^{m^*}}\right)\right)\middle| k \in \{1, ..., r\}\right\}$$

the singular locus. Now write the pairs in this set in the form $(\rho(t), \alpha_k(t) + \delta_k(t))$, where α_k resp. δ_k denotes the meromorphic resp. holomorphic part. Next we cancel redundant pairs, i.e., the pairs $(\rho(t), \alpha_k(t)+\delta_k(t))$ such that ζ with $\zeta^{m^*+n^*} = 1$ and an $l \in \{1, ..., r\} \setminus \{k\}$ with $\alpha_k(\zeta t) + \delta_k(\zeta t) = \alpha_l(t) + \delta_l(t)$ exists. Then we receive a minimal representation (after renumbering the parametrizations, if necessary)

$$\mathscr{S} := \{(\rho(t), \alpha_k(t) + \delta_k(t)) |\, k = 1, ..., r^*\}$$

of the singular locus. With Theorem 3.3.2.4 (noting the fact that it is sufficient to consider only the meromorphic parts α_k),

$$\left(\rho^+\left(\mathrm{El}(\mathrm{Id}, \varphi, R) *_0 \mathrm{El}\left(\mathrm{Id}, \psi, S\right)\right)\right)_0^{\wedge,\mathrm{loc.}} \cong \bigoplus_{\alpha \in \Gamma} \mathscr{E}^\alpha \otimes T_\alpha,$$

where $\Gamma := \left\{\alpha_k(\xi t) \,\middle|\, \xi \in \mathbb{C}^* : \xi^{m^*+n^*} = 1; k = 1, ..., r^*\right\}$. □

Note that the elements $\alpha_k(\xi t)$ with $k \in \{1, ..., r^*\}$ and $\xi^{m^*+n^*} = 1$ do not have to be pairwise non-identical. Therefore we decompose Γ (after a suitable renumbering) in h disjoint subsets, $\Gamma = \biguplus_{k=1}^h \Gamma_k$, with

$$\Gamma_k := \left\{\alpha \in \Gamma \,\middle|\, \exists \xi \in \mathbb{C} : \xi^{m^*+n^*} = 1 \wedge \alpha(\xi t) = \alpha_k(t)\right\}.$$

Clearly $h \leq r^* \leq r \leq m$. According to this decomposition of Γ, we want to prove that the decomposition of the convolution from Theorem 3.3.2.5 can be rewritten as a decomposition of $(\mathrm{El}(\mathrm{Id}, \varphi, R) *_0 \mathrm{El}(\mathrm{Id}, \psi, S))_0^{\wedge,\mathrm{loc.}}$, i.e., we want to show

$$(\mathrm{El}(\mathrm{Id}, \varphi, R) *_0 \mathrm{El}(\mathrm{Id}, \psi, S))_0^{\wedge,\mathrm{loc.}} \cong \bigoplus_{k=1}^{h} \mathrm{El}(\rho, \alpha_k, T_k)$$

for some regular meromorphic connections T_k. So we have to prove $\alpha, \bar{\alpha} \in \Gamma_k \Rightarrow T_\alpha \cong T_{\bar{\alpha}}$ for all $k = 1, ..., h$.

Let $\rho(t) = t^{m^*+n^*}$ and $\alpha(t) \in \Gamma$. T_α can then be found as described in Chapter 2 (Property 3 of the nearby cycles functor) which is the same method as the one used in Theorem 3.2.2 (cf. [6, Section 1.2]): At first, the ramification needed to compute T_α with respect to $\alpha(\eta)$ is $\rho : \eta \mapsto z$. From now on, we work as in the proof of Theorem 3.2.2. which tells us that we have to take the direct image of

$$\mathscr{E}^{\varphi(u)+\psi\left(\frac{\rho(\eta)}{u}\right)} \otimes T$$

by the projection ϕ to $\mathbb{A}_{m,\eta}$. For a better overview, we consider the diagram

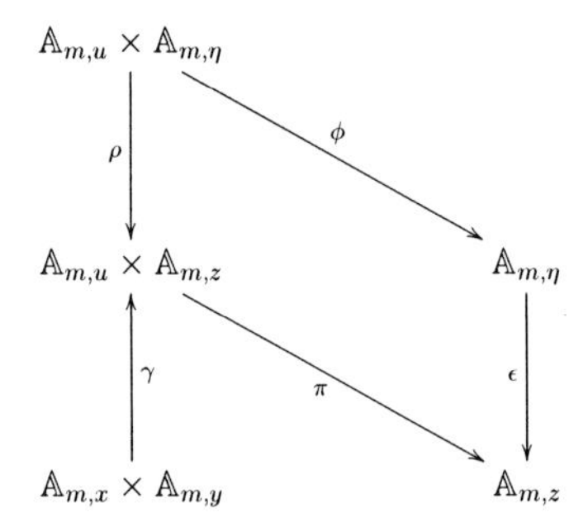

with a morphism $\epsilon : \mathbb{A}_{m,\eta} \to \mathbb{A}_{m,z}$. We started with the module $\mathscr{E}^{\varphi(u)+\psi\left(\frac{z}{u}\right)} \otimes T$ over $\mathbb{A}_{m,u} \times \mathbb{A}_{m,z}$ which we denote by M. In order to determine the regular part with respect to the exponential term $\mathscr{E}^{\alpha(\eta)}$, we have to study the module

$$\Psi_\eta \left(\epsilon^+ \pi_+ M \otimes \mathscr{E}^{-\alpha(\eta)}\right)$$

over $\mathbb{A}_{m,\eta}$. But the proper base change formula [6, Lemma 1.1] tells us

$$\epsilon^+ \pi_+ M = \phi_+ \rho^+ M.$$

It follows from the projection formula [1, p.3]

$$\phi_+ \rho^+ M \otimes \mathscr{E}^{-\alpha(\eta)} \cong \phi_+ \left(\rho^+ M \otimes \phi^+ \mathscr{E}^{-\alpha(\eta)}\right).$$

This gives

$$\Psi_\eta\left(\epsilon^+\pi_+ M \otimes \mathscr{E}^{-\alpha(\eta)}\right) \cong \Psi_\eta\left(\phi_+\left(\rho^+ M \otimes \phi^+\mathscr{E}^{-\alpha(\eta)}\right)\right).$$

Now the functors $\Psi_\eta(.)$ and ϕ_+ commute (see Chapter 2, Property 2). That is

$$\begin{aligned}\Psi_\eta\left(\phi_+\left(\rho^+ M \otimes \phi^+\mathscr{E}^{-\alpha(\eta)}\right)\right) &\cong \phi_+\Psi_\eta\left(\rho^+ M \otimes \phi^+\mathscr{E}^{-\alpha(\eta)}\right)\\ &\cong \phi_+\Psi_\eta\left(\mathscr{E}^{\varphi(u)+\psi\left(\frac{\rho(\eta)}{u}\right)-\alpha(\eta)} \otimes T\right)\end{aligned}$$

because $\phi^+\mathscr{E}^{-\alpha(\eta)} \cong \mathscr{E}^{-\alpha(\eta)}$. This leads us to the computation of

$$\Psi_\eta\left(\mathscr{E}^{\varphi(u)+\psi\left(\frac{\rho(\eta)}{u}\right)-\alpha(\eta)} \otimes T\right).$$

We use the notation

$$\varphi(u) = \sum_{i=1}^{n} a_i \frac{1}{u^i}, \psi\left(\frac{\rho(\eta)}{u}\right) = \sum_{j=1}^{m} b_j \frac{u^j}{\eta^{j(n^*+m^*)}}, \alpha(\eta) = \sum_{l=1}^{m^* n} e_l \frac{1}{\eta^l}$$

from above. Note that $m^* n = mn^*$. We have

$$\begin{aligned}&\varphi(u) + \psi\left(\frac{\rho(\eta)}{u}\right) - \alpha(\eta)\\ &= \frac{1}{u^n \eta^{m(n^*+m^*)}}\left(\eta^{m(n^*+m^*)} a(u) + u^n \sum_{j=1}^{m} b_j u^j \eta^{(m-j)(n^*+m^*)} - u^n \eta^{m^* m} e(\eta)\right)\end{aligned}$$

where $a(u) := \sum_{i=1}^{n} a_i u^{n-i}$ and $e(\eta) := \sum_{l=1}^{m^* n} e_l \eta^{mn^*-l}$ are holomorphic functions.
Next we have to blow up the ideal (u, η) in order to solve the singularity at $(0, 0)$. For this, we need a composition of m^* such blowups. This gives a resolution $\nu = (\nu_1, \nu_2) : \mathbb{X} \to \mathbb{C}^2$. Consider the diagram

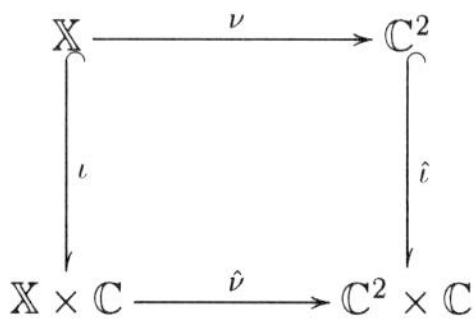

$$\begin{array}{ccc} \mathbb{X} & \xrightarrow{\nu} & \mathbb{C}^2 \\ \Big\downarrow \iota & & \Big\downarrow \hat{\iota} \\ \mathbb{X} \times \mathbb{C} & \xrightarrow{\hat{\nu}} & \mathbb{C}^2 \times \mathbb{C} \end{array}$$

with $\iota = (\mathrm{Id}_{\mathbb{X}}, \nu_2), \hat{\nu} = (\nu, \mathrm{Id}_{\mathbb{C}}), \hat{\iota}(v, \eta) = (v, \eta, \eta)$. Moreover we define the induced maps

$$\begin{aligned}\tilde{\nu} &: \mathbb{X} \times \{0\} \to \mathbb{C}^2 \times \{0\}\\ \tilde{\iota} &: \{0\} \times \mathbb{C} \to \{0\} \times \mathbb{C}^2.\end{aligned}$$

3.3.2.6 Lemma: With the previous setting, the following holds:

1) $\Psi_\eta\left(\mathscr{E}^{\varphi(u)+\psi\left(\frac{\rho(\eta)}{u}\right)-\alpha(\eta)}\otimes T\right)$ is isomorphic to
$\tilde{\nu}_+\Psi_{\eta\circ\nu}\left(\iota_+\nu^+\left(\mathscr{E}^{\varphi(u)+\psi\left(\frac{\rho(\eta)}{u}\right)-\alpha(\eta)}\otimes T\right)\right)$.

2) Let U_i be the chart in the coordinates (v,η) with $e_i:(v,\eta)\mapsto\left(v\eta^{m^*-i},v^i\eta\right)$ for all $i\in\{0,...,m\}$. Then

$$i>0\Rightarrow\Psi_{\eta\circ\nu}\left(\iota_+\nu^+\left(\mathscr{E}^{\varphi(u)+\psi\left(\frac{\rho(\eta)}{u}\right)-\alpha(\eta)}\otimes T\right)\right)\Big|_{U_i}=0.$$

Proof:

1) The statement is just Property 4 (Chapter 2) of the nearby cycles functor. We want to explain this here and use the proof of [7, Lemma 4.5]: Let us define $\mathscr{P}:=\nu^+\left(\mathscr{E}^{\varphi(u)+\psi\left(\frac{\rho(\eta)}{u}\right)-\alpha(\eta)}\otimes T\right)$. Then

$$\begin{aligned}&\Psi_\eta\left(\mathscr{E}^{\varphi(u)+\psi\left(\frac{\rho(\eta)}{u}\right)-\alpha(\eta)}\otimes T\right)\cong\tilde{\iota}_+\Psi_\eta\left(\mathscr{E}^{\varphi(u)+\psi\left(\frac{\rho(\eta)}{u}\right)-\alpha(\eta)}\otimes T\right)\\&\cong\tilde{\iota}_+\Psi_\eta\left(\nu_+\nu^+\left(\mathscr{E}^{\varphi(u)+\psi\left(\frac{\rho(\eta)}{u}\right)-\alpha(\eta)}\otimes T\right)\right)\cong\Psi_{\mathbb{C}^2\times\{0\}}\left(\hat{\iota}_+\nu_+\mathscr{P}\right)\\&\cong\Psi_{\mathbb{C}^2\times\{0\}}\left(\hat{\nu}_+\iota_+\mathscr{P}\right)\cong\tilde{\nu}_+\Psi_{\mathbb{X}\times\{0\}}\left(\iota_+\mathscr{P}\right)=\tilde{\nu}_+\Psi_{\eta\circ\nu_2}\left(\iota_+\mathscr{P}\right).\end{aligned}$$

The first isomorphism is clear due to Kashiwara's equivalence because the map $\tilde{\iota}:\{0\}\times\mathbb{C}\to\{0\}\times\mathbb{C}^2$ is induced by $\hat{\iota}$ and maps $(0,\eta)$ to $(0,\eta,\eta)$. The second isomorphism holds due to [7, Lemma 4.4], the third one holds because the nearby cycle functor commutes with proper direct images, and the fourth one holds due to the fact $\hat{\iota}_+\nu_+=(\hat{\iota}\circ\nu)_+=(\hat{\nu}\circ\iota)_+=\hat{\nu}_+\iota_+$. The last isomorphism is the commutation property of the nearby cycle functor with proper direct images again. This proves 1).

2) Consider

$$\begin{aligned}\mathscr{P}&=\nu^+\left(\mathscr{E}^{\varphi(u)+\psi\left(\frac{\rho(\eta)}{u}\right)-\alpha(\eta)}\otimes T\right)\Big|_{U_i}\\&=e_i^+\left(\mathscr{E}^{\varphi(u)+\psi\left(\frac{\rho(\eta)}{u}\right)-\alpha(\eta)}\otimes T\right)\\&=\mathscr{E}^{\varphi\left(v\eta^{m^*-i}\right)+\psi\left(\frac{\rho\left(v^i\eta\right)}{v\eta^{m^*-i}}\right)-\alpha(v^i\eta)}\otimes e_i^+T.\end{aligned}$$

We have

$$\varphi\left(v\eta^{m^*-i}\right)+\psi\left(\frac{\rho\left(v^i\eta\right)}{v\eta^{m^*-i}}\right)-\alpha\left(v^i\eta\right)=\frac{1}{v^{n+im(n^*+m^*)}\eta^{2nm^*+mm^*-ni}}f_i(v,\eta)$$

with

$$\begin{aligned}f_i(v,\eta)&=v^{im(n^*+m^*)}\eta^{m(n^*+m^*)}a\left(v\eta^{m^*-i}\right)\\&+v^n\eta^{n(m^*-i)}\sum_{j=1}^{m}b_jv^j\eta^{j(m^*-i)}\left(v^i\eta\right)^{(m-j)(n^*+m^*)}\\&+v^n\eta^{n(m^*-i)}v^{imm^*}\eta^{mm^*}e\left(v^i\eta\right)\\&=v^{n+m}\eta^{(m^*-i)(n+m)}g_i(v,\eta)\end{aligned}$$

(where a and e are the holomorphic functions from page 34) and

$$\begin{aligned}g_i(v,\eta)&=v^{(im^*-1)(n+m)}\eta^{i(n+m)}a\left(v\eta^{m^*-i}\right)\\&+\sum_{j=1}^{m}b_jv^{(m-j)(in^*+im^*-1)}\eta^{(n^*+i)(m-j)}\\&+v^{m(im^*-1)}\eta^{im}e\left(v^i\eta\right).\end{aligned}$$

Hence

$$\varphi\left(v\eta^{m^*-i}\right)+\psi\left(\frac{\rho\left(v^i\eta\right)}{v\eta^{m^*-i}}\right)-\alpha(v^i\eta)=\frac{1}{v^{m(in^*+im^*-1)}\eta^{m(n^*+i)}}g_i(v,\eta).$$

Now let $i>0$. Remember $\iota:\mathbb{X}\to\mathbb{X}\times\mathbb{C},(v,\eta)\mapsto(v,\eta,\nu_2)$. On U_i, ν_2 acts on η just like e_i does, hence $\nu_2(\eta)=v^i\eta$.

$$\Psi_{\eta\circ\nu_2}\left(\iota_+\nu^+\left(\mathscr{E}^{\varphi(u)+\psi\left(\frac{\rho(\eta)}{u}\right)-\alpha(\eta)}\otimes T\right)\right)\Big|_{U_i}=\Psi_{\eta\circ\nu_2}\left(\iota_+\left(\mathscr{E}^{g_i(v,\eta)/v^k\eta^l}\otimes e_i^+T\right)\right)$$

with $k:=m(in^*+im^*-1)\geq 1$ and $l:=m(n^*+i)\geq 1$. Now we use Chapter 2, in particular Property 5 of the nearby cycle functor:
By definition, the module $\Psi_{\eta\circ\nu_2}\left(e_i^+T\right)$ is supported on $\{v^i\eta=0\}$. Moreover it is even supported on $\{\eta=0\}$ (see the setting of Prop. 2.1). But at every point, we can change coordinates in order to get the situation in [7, Lemma 7.3]. Hence the Lemma follows. □

Lemma 3.3.2.6 shows that the only chart to consider is U_0 with $e_0 : (v, \eta) \mapsto (v\eta^{m^*}, \eta)$. Then the module to consider is of the form

$$\Psi_\eta \left(\iota_+ \left(\mathscr{E}^{\varphi\left(v\eta^{m^*}\right)+\psi\left(\frac{\rho(\eta)}{v\eta^{m^*}}\right)-\alpha(\eta)} \otimes e_0^+ T \right) \right).$$

ι is a proper map on U_0, so it commutes with Ψ_η. Therefore we have to compute

$$\Psi_\eta \left(\mathscr{E}^{\varphi\left(v\eta^{m^*}\right)+\psi\left(\frac{\rho(\eta)}{v\eta^{m^*}}\right)-\alpha(\eta)} \otimes e_0^+ T \right).$$

We have

$$\begin{aligned}
&\varphi\left(v\eta^{m^*}\right) + \psi\left(\frac{\rho(\eta)}{v\eta^{m^*}}\right) - \alpha(\eta) \\
=&\frac{1}{v^n \eta^{m(2n^*+m^*)}} \left(\eta^{m(n^*+m^*)} \left(a\left(v\eta^{m^*}\right) + v^n \sum_{j=1}^{m} b_j v^j \eta^{n^*(m-j)} - v^n e(\eta) \right)\right) \\
=&\frac{1}{v^n \eta^{nm^*}} \left(a\left(v\eta^{m^*}\right) + v^n \sum_{j=1}^{m} b_j v^j \eta^{n^*(m-j)} - v^n e(\eta) \right) =: \frac{f_\alpha(v,\eta)}{v^n \eta^{nm^*}}.
\end{aligned}$$

So equivalently, we can compute

$$\Psi_\eta \left(\mathscr{E}^{f_\alpha(v,\eta)/v^n\eta^{nm^*}} \otimes e_0^+ T \right). \tag{3.2}$$

Due to Lemma 3.2.3, we are only interested in values $\bar{v}$ with $(v-\bar{v})^2 | f_\alpha(v, \eta)$. But values $\bar{v}$ with that property especially have to be double roots of $f_\alpha(v, 0)$. In particular, we know that the module (3.2) is supported on $\eta = 0$ by definition and moreover at most on the set $\{f_\alpha(v, 0) = 0\}$. So let us first consider $f_\alpha(v, 0)$. We have

$$f_\alpha(v, 0) = a_n + v^{n+m} b_m + v^n e_{m^* n}.$$

Considering

$$\frac{\partial^2 f_\alpha}{\partial v^2}(v, 0) = (m+n)(m+n-1) b_m v^{m+n-2} - n(n-1) e_{m^* n} v^{n-2}$$

in the case $n \geq 2$, it is easy to see that $f_\alpha(v, 0)$ has no roots of order > 2 (but $v = 0$ if $n > 2$; however, $v = 0$ is of no interest for us). Hence all roots of $f_\alpha(v, 0)$ have order ≤ 2. As $f_\alpha(v, 0)$ has no roots $\hat{v}$ of order > 2, $(v - \hat{v})^b | f_\alpha(v, \eta)$ for $b > 2$ is not possible, too. Next we give an important statement concerning the values $\bar{v}$ with $(v - \bar{v})^2 | f_\alpha(v, \eta)$.

Proposition 3.3.2.7: Let $\alpha \in \Gamma$ and $f_{\alpha(\kappa\eta)}(v,\eta)$ be the corresponding polynomial of $\alpha(\kappa\eta)$. Then for all κ with $\kappa^{m^*+n^*} = 1$, the number of values $\hat{v}$ with the property $(v-\hat{v})^2 | f_{\alpha(\kappa\eta)}(v,\eta)$ is equal.

Proof: Changing $\alpha(\eta)$ to $\alpha(\kappa\eta) = \sum_{l=1}^{m^*n} e_l \kappa^{-l}\eta^{-l}$ leads to the computation of the module

$$\Psi_\eta \left(\mathscr{E}^{\varphi(u)+\psi\left(\frac{\rho(\eta)}{u}\right)-\alpha(\kappa\eta)} \otimes T \right)$$

with

$$h_{\alpha(\kappa\eta)}(u,\eta) := \varphi(u) + \psi\left(\frac{\rho(\eta)}{u}\right) - \alpha(\kappa\eta).$$

Due to $\rho(\eta) = \eta^{m^*+n^*}$, we have $\rho(\eta) = \rho(\kappa\eta)$. This gives

$$h_{\alpha(\kappa\eta)}(u,\eta) = h_{\alpha(\eta)}(u,\kappa\eta). \tag{3.3}$$

Next we have to apply the inverse image e_0^+ under the chart with coordinates (v,η) with $e_0 : (v,\eta) \mapsto \left(v\eta^{m^*}, \eta\right)$. We define $\tilde{h}_{\alpha(\kappa\eta)}(v,\eta) := h_{\alpha(\kappa\eta)}\left(v\eta^{m^*}, \eta\right)$ for all κ with $\kappa^{m^*+n^*} = 1$. Then with (3.3) and $\kappa^{-m^*} = \kappa^{n^*}$ $\left(\text{because of } \kappa^{m^*+n^*} = 1\right)$,

$$\begin{aligned} \tilde{h}_{\alpha(\kappa\eta)}(v,\eta) &= h_{\alpha(\kappa\eta)}\left(v\eta^{m^*},\eta\right) = h_{\alpha(\eta)}\left(v\eta^{m^*},\kappa\eta\right) = h_{\alpha(\eta)}\left(\left(\kappa^{-m^*}v\right)(\kappa\eta)^{m^*},\kappa\eta\right) \\ &= \tilde{h}_{\alpha(\eta)}\left(\kappa^{-m^*}v,\kappa\eta\right) = \tilde{h}_{\alpha(\eta)}\left(\kappa^{n^*}v,\kappa\eta\right). \end{aligned}$$

This means, because of $\tilde{h}_{\alpha(\eta)}(v,\eta) = \frac{f_{\alpha(\eta)}(v,\eta)}{v^n\eta^{nm^*}}$,

$$\begin{aligned} f_{\alpha(\kappa\eta)}(v,\eta) &= v^n\eta^{nm^*} \cdot \tilde{h}_{\alpha(\kappa\eta)}(v,\eta) = v^n\eta^{nm^*} \cdot \tilde{h}_{\alpha(\eta)}\left(\kappa^{m^*}v,\kappa\eta\right) \\ &= \left(\kappa^{n^*}v\right)^n(\kappa\eta)^{nm^*} \cdot \tilde{h}_{\alpha(\eta)}\left(\kappa^{m^*}v,\kappa\eta\right) = f_{\alpha(\eta)}\left(\kappa^{n^*}v,\kappa\eta\right). \end{aligned}$$

Finally let $\hat{v}$ be a value with $f_{\alpha(\eta)}(v,\eta) = (v-\hat{v})^2 \cdot \tilde{f}_{\alpha(\eta)}(v,\eta)$. Then

$$\begin{aligned} f_{\alpha(\kappa\eta)}(v,\eta) &= f_{\alpha(\eta)}\left(\kappa^{n^*}v,\kappa\eta\right) = \left(\kappa^{n^*}v-\hat{v}\right)^2 \cdot \tilde{f}_{\alpha(\eta)}\left(\kappa^{n^*}v,\kappa\eta\right) \\ &= \left(v-\kappa^{m^*}\hat{v}\right)^2\kappa^{2n^*} \cdot \tilde{f}_{\alpha(\eta)}\left(\kappa^{n^*}v,\kappa\eta\right). \end{aligned}$$

This proves the lemma. □

Now, like in the proof of Theorem 3.2.2, $\frac{f_{\alpha(\kappa\eta)}(v,\eta)}{v^n \eta^{m*n}}$ can be written as $\frac{g_{\alpha(\kappa\eta)}(v',\eta)}{\eta^{m*n}}$, where $(v',\eta) = (v - \hat{v}, \eta)$ (with a root $\hat{v}$ of $f_{\alpha(\kappa\eta)}(v,0)$) are suitable local coordinates and

$$g_{\alpha(\kappa\eta)}(v',\eta) = \begin{cases} v', & (v-\hat{v}) \mid f_{\alpha(\kappa\eta)}(v,\eta), \ (v-\hat{v})^2 \nmid f_{\alpha(\kappa\eta)}(v,\eta) \\ v'^2, & (v-\hat{v})^2 \mid f_{\alpha(\kappa\eta)}(v,\eta). \end{cases}$$

Then we get the following:

3.3.2.8 Corollary: Let $\Gamma = \biguplus_{k=1}^{h} \Gamma_k$ be defined as on page 31f. Moreover let

$$\left(\rho^+ \left(\mathrm{El}(\mathrm{Id}, \varphi, R) *_0 \mathrm{El}\left(\mathrm{Id}, \psi, S\right)\right)\right)_0^{\wedge,\mathrm{loc.}} \cong \bigoplus_{\alpha\in\Gamma} \mathscr{E}^{\alpha} \otimes T_\alpha$$

be the decomposition with respect to Theorem 3.3.2.4. Then for all $k = 1, ..., h$ and all $\alpha \in \Gamma_k$, the corresponding regular parts T_α are isomorphic.

Proof: Let us consider the functions $\tilde{h}_{\alpha(\kappa\eta)}(v,\eta) = \frac{f_{\alpha(\kappa\eta)}(v,\eta)}{v^n \eta^{nm^*}}$ which we used in the proof of Proposition 3.3.2.7. Then, due to the decomposition of Th. 3.3.2.4 and [5, Ex. 5.2.1],

$$\Psi_\eta \left(\mathscr{E}^{\tilde{h}_{\alpha(\eta)}(v,\eta)} \otimes e_0^+ T\right) \cong \Psi_\eta \left(T_{\alpha(\eta)}\right) \text{ and } \Psi_\eta \left(\mathscr{E}^{\tilde{h}_{\alpha(\kappa\eta)}(v,\eta)} \otimes e_0^+ T\right) \cong \Psi_\eta \left(T_{\alpha(\kappa\eta)}\right).$$

As $\omega : (v,\eta) \mapsto \left(\kappa^{n^*} v, \kappa\eta\right)$ is a proper automorphism, we get the local isomorphisms

$$\begin{aligned} \Psi_\eta \left(T_{\alpha(\eta)}\right) &\cong \Psi_\eta \left(\mathscr{E}^{\tilde{h}_{\alpha(\eta)}(v,\eta)} \otimes e_0^+ T\right) \cong \Psi_\eta \left(\mathscr{E}^{\tilde{h}_{\alpha(\eta)}\left(\kappa^{n^*} v, \kappa\eta\right)} \otimes e_0^+ T\right) \\ &= \Psi_\eta \left(\mathscr{E}^{\tilde{h}_{\alpha(\kappa\eta)}(v,\eta)} \otimes e_0^+ T\right) \cong \Psi_\eta \left(T_{\alpha(\kappa\eta)}\right) \end{aligned}$$

with the help of Proposition 3.3.2.7. □

As a consequence, we can write the decomposition

$$\left(\rho^+ \left(\mathrm{El}(\mathrm{Id}, \varphi, R) *_0 \mathrm{El}\left(\mathrm{Id}, \psi, S\right)\right)\right)_0^{\wedge,\mathrm{loc.}} \cong \bigoplus_{\alpha\in\Gamma} \mathscr{E}^{\alpha} \otimes T_\alpha$$

in the desired form

$$\left(\mathrm{El}(\mathrm{Id}, \varphi, R) *_0 \mathrm{El}\left(\mathrm{Id}, \psi, S\right)\right)_0^{\wedge,\mathrm{loc.}} \cong \bigoplus_{k=1}^{h} \mathrm{El}\left(\rho, \alpha_k, T_{\alpha_k}\right)$$

with $\alpha_k \in \Gamma_k$ for every $k = 1, ..., h$.

Now the last thing for us to show is the fact that the regular parts T_{α_k} are not zero. Note that this is already a statement of Theorem 3.3.2.4, but we also want to show that there really is a possibility to prove the existence of a value $\bar{v}$ with

$$\Psi_\eta\left(\mathscr{E}^{(v-\bar{v})^2/\eta^{nm^*}} \otimes e_0^+ T\right) \neq 0.$$

3.3.2.9 Lemma: Consider the decomposition

$$(\mathrm{El}(\mathrm{Id}, \varphi, R) *_0 \mathrm{El}(\mathrm{Id}, \psi, S))_0^{\wedge,\mathrm{loc.}} \cong \bigoplus_{k=1}^{h} \mathrm{El}(\rho, \alpha_k, T_{\alpha_k}).$$

from above. Then for all k, $\mathrm{El}(\rho, \alpha_k, T_{\alpha_k}) \neq 0$.

Proof: Let us start in the situation on page 30. We used power series $\lambda_k(t) \in t\mathbb{C}[\![t]\!]$ with $\lambda_k'(0) \neq 0$ and $\tilde{\rho}_k(\lambda_k(t)) = t^{m^*+n^*}$. Due to [1, Lemma 2.2], we have the following equivalence for automorphisms μ on the formal disc

$$\mathrm{El}(\rho, \alpha_k, T_{\alpha_k}) \neq 0 \Leftrightarrow \mathrm{El}(\rho \circ \mu, \alpha_k \circ \mu, T_{\alpha_k \circ \mu}) \neq 0 \Leftrightarrow T_{\alpha_k \circ \mu} \neq 0.$$

So, as $\lambda_k(t)$ is such an automorphism, we just reverse it and work with

$$\tilde{\rho}_k(t) = z_k\left(t^{m^*}\right), \tilde{\alpha}_k(t) + \tilde{\delta}_k(t) = \varphi\left(t^{m^*}\right) + \psi\left(\frac{z_k\left(t^{m^*}\right)}{t^{m^*}}\right)$$

again (just as on page 30). So we have to compute

$$\Psi_\eta\left(\mathscr{E}^{\varphi(u)+\psi\left(\frac{\tilde{\rho}_k(\eta)}{u}\right)-\tilde{\alpha}_k(\eta)-\tilde{\delta}_k(\eta)} \otimes T\right).$$

As the computation works in consideration of the holomorphic term $\tilde{\delta}_k(\eta)$, we include it.

$$\begin{aligned}
&\varphi(u) + \psi\left(\frac{\tilde{\rho}_k(\eta)}{u}\right) - \tilde{\alpha}_k(\eta) - \tilde{\delta}_k(\eta) = \varphi(u) + \psi\left(\frac{\tilde{\rho}_k(\eta)}{u}\right) - \varphi\left(\eta^{m^*}\right) - \psi\left(\frac{\tilde{\rho}_k(\eta)}{\eta^{m^*}}\right) \\
=&\sum_{i=1}^{n} a_i\left(u^{-i} - \eta^{im^*}\right) + \sum_{j=1}^{m} b_j\left(\frac{u^j}{\tilde{\rho}_k(\eta)^j} - \frac{\eta^{jm^*}}{\tilde{\rho}_k(\eta)^j}\right) \\
=&\frac{1}{u^n\eta^{m(m^*+n^*)}}\sum_{i=1}^{n} a_i\left(u^{n-i}\eta^{m(m^*+n^*)} - u^n\eta^{mm^*+m^*(n-i)}\right) \\
+&\frac{1}{u^n\eta^{m(m^*+n^*)}}\sum_{j=1}^{m} b_j u^n \eta^{(m-j)(n^*+m^*)}\left(\frac{u^j}{\bar{\rho}_k(\eta)^j} - \frac{\eta^{jm^*}}{\bar{\rho}_k(\eta)^j}\right)
\end{aligned}$$

with $\bar{\rho}_k(\eta) := \frac{\tilde{\rho}_k(\eta)}{\eta^{m^*+n^*}}$. Note that $\bar{\rho}_k(0) \neq 0$ because $\tilde{\rho}_k(\eta)$ has order $m^* + n^*$. Next we consider the chart $e_0 : (v, \eta) \mapsto \left(v\eta^{m^*}, \eta\right)$ again (see Lemma 3.3.2.6). Due to $m^*n = mn^*$,

$$
\begin{aligned}
&\frac{1}{v^n\eta^{mm^*+2mn^*}} \sum_{i=1}^{n} a_i \left(v^{n-i}\eta^{mm^*+2mn^*-im^*} - v^n\eta^{2mn^*+mm^*-im^*}\right) \\
&+\frac{1}{v^n\eta^{mm^*+2mn^*}} \sum_{j=1}^{m} b_j \left(\frac{v^{n+j}\eta^{2mn^*+mm^*-jn^*}}{\bar{\rho}_k(\eta)^j} - \frac{v^n\eta^{2mn^*+mm^*-jn^*}}{\bar{\rho}_k(\eta)^j}\right) \\
&=\frac{1}{v^n\eta^{nm^*}} \left(\sum_{i=1}^{n} a_i\eta^{m^*(n-i)} \left(v^{n-i} - v^n\right) + \sum_{j=1}^{m} b_j v^n \eta^{n^*(m-j)} \left(\frac{v^j}{\bar{\rho}_k(\eta)^j} - \frac{1}{\bar{\rho}_k(\eta)^j}\right)\right) \\
&=: \frac{\tilde{f}_k(v,\eta)}{v^n\eta^{nm^*}}.
\end{aligned}
$$

So again, it is equivalent to compute $\Psi_{\eta=0}\left(\mathscr{E}^{\tilde{f}_k(v,\eta)/v^n\eta^{nm^*}} \otimes e_0^+T\right)$. Now it is easy to see that $\tilde{f}_k(1,\eta) = 0$. Moreover

$$
\begin{aligned}
\frac{\partial \tilde{f}_k}{\partial v}(v,\eta) &= -nv^{n-1}a_n + \sum_{i=1}^{n-1} a_i\eta^{m^*(n-i)} \left((n-i)v^{n-i-1} + nv^{n-1}\right) \\
&+ \sum_{j=1}^{m} b_j\eta^{n^*(m-j)} \left(\frac{(n+j)v^{n+j-1}}{\bar{\rho}_k(\eta)^j} - \frac{nv^{n-1}}{\bar{\rho}_k(\eta)^j}\right).
\end{aligned}
$$

Hence, replacing $\bar{\rho}_k$ by $\tilde{\rho}_k$ again,

$$
\begin{aligned}
\frac{\partial \tilde{f}_k}{\partial v}(1,\eta) &= -\sum_{i=1}^{n} ia_i\eta^{m^*(n-i)} + \sum_{j=1}^{m} jb_j \frac{\eta^{n^*(m-j)}}{\bar{\rho}_k(\eta)^j} \\
&= -\sum_{i=1}^{n} ia_i\eta^{m^*(n-i)} + \sum_{j=1}^{m} jb_j \frac{\eta^{j(n^*+m^*)}\eta^{n^*(m-j)}}{\eta^{j(m^*+n^*)}\tilde{\rho}(\eta)^j} \\
&= \eta^{m^*n} \left(-\sum_{i=1}^{n} ia_i \left(\eta^{m^*}\right)^{-i} + \sum_{j=1}^{m} jb_j \left(\frac{\eta^{m^*}}{\tilde{\rho}_k(\eta)}\right)^j\right) = 0.
\end{aligned}
$$

The equation equals zero because the term in brackets is just the defining equation of C' (see page 26) - and the pair $\left(\eta^{m^*}, \tilde{\rho}_k(\eta)\right)$ a corresponding solution. Another computation gives $\frac{\partial^2 \tilde{f}_k}{\partial v^2}(1,\eta) \neq 0$. As final result, $(v-1)^2$ divides $\tilde{f}_k(v,\eta)$. Therefore, in a neighbourhood of the point $(v,\eta) = (1,0)$, the function $\tilde{f}_k(v,\eta)$ can be written as $\tilde{f}_k(v,\eta) = (v-1)^2 \cdot$ unit. So we can apply Lemma 3.2.3 which shows that the regular part $T_{\tilde{\alpha}_k}$ is not equal to zero, hence $T_\alpha \neq 0$. □

Now we are able to formulate the main statement of this chapter which finally gives the desired formula for the local formal convolution in the multiplicative case.

3.3.2.10 Theorem: Let $\varphi(x) = \sum_{i=1}^{n} a_i x^{-i}$ resp. $\psi(y) = \sum_{j=1}^{m} b_j y^{-j}$ be meromorphic functions with pole order n resp. m (i.e. $a_n \neq 0$ resp. $b_m \neq 0$). Let $d := \gcd(n, m)$, $n^* := \frac{n}{d}$ and $m^* := \frac{m}{d}$. Then

(I) Applying the direct image γ_+ with $\gamma : (x, y) \mapsto (x, xy) =: (u, z)$ to the set $C = \{x\varphi'(x) = y\psi'(y)\}$, we receive C' (see page 26). Then the r pairwise distinct irreducible components of the defining polynomial equation of the set C' have parametrizations of the form

$$\left(t^{m^*}, \sum_{p=1}^{\infty} c_{p,k} t^{m^*+n^*+p-1}\right) =: \left(t^{m^*}, \tilde{\rho}_k(t)\right),\ k \in \{1, ..., r\}.$$

(II) For all $k \in \{1, ..., r\}$, we define

$$\varphi\left(t^{m^*}\right) + \psi\left(\frac{\tilde{\rho}_k(t)}{t^{m^*}}\right) = \sum_{l=1}^{m^* n} d_{l,k} t^{-l} + \sum_{l=0}^{\infty} \tilde{d}_{l,k} t^{l} =: \tilde{\alpha}_k(t) + \tilde{\delta}_k(t).$$

(III) Let $\lambda_k(t) \in t\mathbb{C}[\![t]\!]$ with $\lambda_k'(0) \neq 0$ and $(\tilde{\rho}_k \circ \lambda_k)(t) = t^{m^*+n^*}$. We denote by $\alpha_k(t)$ resp. $\delta_k(t)$ the meromorphic resp. holomorphic part of $(\tilde{\alpha}_k + \tilde{\delta}_k) \circ \lambda_k$ and cancel redundant pairs $(\rho, \alpha_k + \delta_k)$ in the sense of Theorem 3.3.2.5. Afterwards we omit the holomorphic parts and receive the minimal set of parametrizations

$$\{(\rho, \alpha_k) | k = 1, ..., r^*\}.$$

With the help of the meromorphic parts $\alpha_k, k \in \{1, ..., r^*\}$, we define the set

$$\Gamma = \biguplus_{k=1}^{h} \Gamma_k = \biguplus_{k=1}^{h} \left\{\alpha \in \Gamma | \exists \xi \in \mathbb{C} : \xi^{m^*+n^*} = 1 \wedge \alpha(\xi t) = \alpha_k(t)\right\}$$

(see Th. 3.3.2.5). Then we get the decomposition

$$(\mathrm{El}(\mathrm{Id}, \varphi, R) *_0 \mathrm{El}(\mathrm{Id}, \psi, S))_0^{\wedge,\mathrm{loc.}} \cong \bigoplus_{k=1}^{h} \mathrm{El}\left(\left[t \mapsto t^{m^*+n^*}\right], \alpha_k, T_{\alpha_k}\right)$$

with regular connections T_{α_k}. □

3.3.2.11 Corollary: Consider the setting of Theorem 3.3.2.10 and let $\gcd(m,n) = 1$. Then the polynomial equation which defines C' has only one irreducible component (see Remark 3.3.2.2). Let its parametrization be given by

$$\left(t^m, \sum_{p=1}^{\infty} c_p t^{m+n+p-1}\right) =: (t^m, \tilde{\rho}(t)).$$

Then we denote by $\tilde{\alpha}(t)$ resp. $\tilde{\delta}(t)$ the meromorphic resp. holomorphic part of the function

$$\varphi(t^m) + \psi\left(\frac{\tilde{\rho}_k(t)}{t^m}\right).$$

Finally let $\lambda(t) \in t\mathbb{C}[\![t]\!]$ with $\lambda'(0) \neq 0$ and $(\tilde{\rho} \circ \lambda)(t) = t^{m+n}$. We denote by $\alpha(t)$ resp. $\delta(t)$ the meromorphic resp. holomorphic part of $(\tilde{\alpha} + \tilde{\delta}) \circ \lambda$. Then it follows

$$(\mathrm{El}(\mathrm{Id}, \varphi, R) *_0 \mathrm{El}(\mathrm{Id}, \psi, S))_0^{\wedge,\mathrm{loc.}} \cong \mathrm{El}\left(\left[t \mapsto t^{m+n}\right], \alpha, T\right)$$

with a regular meromorphic connection T. □

3.3.2.12 Remark: In the setting of Theorem 3.3.2.10, we have the equivalent decomposition

$$(\mathrm{El}(\mathrm{Id}, \varphi, R) *_0 \mathrm{El}(\mathrm{Id}, \psi, S))_0^{\wedge,\mathrm{loc.}} \cong \bigoplus_{k=1}^{h} \mathrm{El}(\tilde{\rho}_k, \tilde{\alpha}_k, T_{\alpha_k})$$

due to [1, Lemma 2.2].

3.3.2.13 Example: Let

$$\varphi(x) = \frac{1}{x} + \frac{1}{x^2} \text{ and } \psi(y) = \frac{1}{y} + \frac{1}{y^2}.$$

So $n = m = 2$. We want to compute

$$(\mathrm{El}([x \mapsto x], \varphi(x), R) *_0 \mathrm{El}([y \mapsto y], \psi(y), S))_0^{\wedge,\mathrm{loc.}}.$$

After Proposition 3.3.1.2,

$$C := \{(x,y) | x\varphi'(x) - y\psi'(y) = 0\} = \left\{(x,y) \middle| -\frac{1}{x} - \frac{2}{x^2} + \frac{1}{y} + \frac{2}{y^2} = 0\right\}.$$

Applying γ_+ with $\gamma : (x,y) \mapsto (x, xy) =: (u,z)$ leads to the set

$$C' := \left\{(u,z) \middle| -\frac{1}{u} - \frac{2}{u^2} + \frac{u}{z} + \frac{u^2}{z^2} = 0\right\}.$$

Multiplying the equation with u^2z^2 then gives the polynomial expression

$$P(u,z) := 2u^4 + u^3z - (2+u)z^2 = 0$$

which we want to parameterize now. It is $m = n = 2$, so after Remark 3.3.2.2, $P(u,z)$ consists of 2 irreducible components. The algorithm in the proof of Lemma 3.3.2.1 tells us that we have to consider the Newton polygon with edges (j, s_j), where j denotes the exponent of z in the above equation, and s_j denotes the lowest u-exponent in the coefficient of z^j. So the edges are given by the points $(0,4),(1,3),(2,0)$. As the polygon with these edges is not convex, the Newton polygon has the single slope 2 different from 0 and ∞. Hence the parametrizations have the form $z(u) = c_1u^2 + \tilde{z}_1$ with $\tilde{z}_1 = \tilde{z}_1(u) \in \mathbb{C}[\![u]\!]$ due to Lemma 3.3.2.1. In order to determine c_1, we consider $P\left(u, c_1u^2 + \tilde{z}_1\right) = 0$ and choose c_1 such that the term with the lowest u-exponent in the coefficient of $\tilde{z}_1^0$ vanishes:

$$P\left(u, c_1u^2 + \tilde{z}_1\right) = \left(2 - 2c_1^2\right)u^4 + \left(c_1 - c_1^2\right)u^5 + \left(u^3 - 4c_1u^2 - 2c_1u^3\right)\tilde{z}_1 - (2+u)\,\tilde{z}_1^2$$

As the term $\left(2 - 2c_1^2\right)u^4$ has to vanish, we get $c_1 = \pm 1$. This gives

$$z_k(u) = (-1)^k u^2 + \tilde{z}_{1,k}, k \in \{1,2\}.$$

Next the same steps have to be repeated with the equations $P_1\left(u, \tilde{z}_{1,k}\right) = 0$, where

$$P_1\left(u, \tilde{z}_{1,k}\right) := P\left(u, (-1)^k u^2 + \tilde{z}_{1,k}\right).$$

Carrying out some iterations of the algorithm, this gives the two parametrizations

$$\begin{aligned} z_1(u) &= -u^2 + \frac{1}{2}u^3 - \frac{1}{4}u^4 \pm \ldots = -u^2\sum_{i=0}^{\infty}\left(-\frac{u}{2}\right)^i = -\frac{2u^2}{2+u}, \\ z_2(u) &= u^2. \end{aligned}$$

As both parametrizations already have integer exponents, we do not have to apply an extra transform $\beta : t \mapsto t^{m^*}$ which we used on page 29. Hence the needed parametrizations are given by the pairs $(u, z_k(u)), k \in \{1,2\}$. Next we define $\tilde{\rho}_k(u) := z_k(u)$. In order to complete the parametrization of the singular locus, we have to compute

$$\tilde{\alpha}_k(u) := \varphi(u) + \psi\left(\frac{\rho_k(u)}{u}\right) \operatorname{mod} \mathbb{C}[\![u]\!].$$

Hence, as there are no redundant parametrizations in the sense of Theorem 3.3.2.5, the minimal set of parametrizations is given by the pairs

$$(\tilde{\rho}_1(u), \tilde{\alpha}_1(u)) = \left(-\frac{2u^2}{2+u}, \frac{2}{u^2} - \frac{1}{u}\right), \ (\tilde{\rho}_2(u), \tilde{\alpha}_2(u)) = \left(u^2, \frac{2}{u^2} + \frac{2}{u}\right).$$

Next we want to find a $\lambda_1(u) \in u\mathbb{C}[\![u]\!]$ with $\lambda_1'(0) \neq 0$ and $\tilde{\rho}_1(\lambda_1(u)) = u^2$. Note that we just have to compute the first three summands of λ_1 because we only need the meromorphic part of the function $\tilde{\alpha}_1(\lambda_1(u))$. One possible solution is

$$\lambda_1(u) = iu - \frac{1}{4}u^2 - \frac{i}{32}u^3 + ...$$

This gives

$$\alpha_1(u) := \tilde{\alpha}_1(\lambda_1(u)) \bmod \mathbb{C}[\![u]\!] = -\frac{2}{u^2}.$$

As $\tilde{\rho}_2$ is already in the right form (i.e., it is a function of the form $u^k, k \in \mathbb{N}$), we set $(\rho, \alpha_2) := (\tilde{\rho}_2(u), \tilde{\alpha}_2(u))$. So we can work with the curves $(\rho, \alpha_k), k \in \{1, 2\}$, with $\rho(u) = u^2$. Theorem 3.3.2.4 gives

$$\rho^+ \left(\mathrm{El}\left([x \mapsto x], \varphi(x), R\right) *_0 \mathrm{El}\left([y \mapsto y], \psi(y), S\right)\right)_0^{\wedge,\mathrm{loc.}} \cong \bigoplus_{\alpha \in \Gamma} \mathscr{E}^{\alpha} \otimes T_{\alpha}$$

with $\Gamma := \{\alpha_k(\xi u) | k = 1, 2; \xi^2 = 1\}$. As $\alpha_1(u) \neq \alpha_2(\xi u)$, we get the decomposition

$$\Gamma = \biguplus_{k=1}^{2} \Gamma_k, \ \Gamma_k := \{\alpha_k(u), \alpha_k(-u)\}.$$

So it remains to check the regular parts T_{α_k} with respect to (ρ, α_k). This means, for $k = 2$ we have to consider

$$\Psi_\eta \left(\mathscr{E}^{\varphi(u) + \psi\left(\frac{\rho(\eta)}{u}\right) - \alpha_2(\eta)} \otimes (R \boxtimes S)\right).$$

We have

$$\begin{aligned} \varphi(u) + \psi\left(\frac{\rho(\eta)}{u}\right) - \alpha_2(\eta) &= \frac{1}{u} + \frac{1}{u^2} + \frac{u}{\eta^2} + \frac{u^2}{\eta^4} - \frac{2}{\eta} - \frac{2}{\eta^2} \\ &= \frac{1}{u^2\eta^4}\left(u\eta^4 + \eta^4 + u^3\eta^2 + u^4 - 2u^2\eta^3 - 2u^2\eta^2\right). \end{aligned}$$

Next we have to solve the singularity at $(u, \eta) = (0, 0)$ by blowing up the ideal

(v,η) with the help of the chart $e_0 : (v,\eta) \mapsto (v\eta, \eta)$. That is, we apply e_0^+. Note that due to Lemma 3.3.2.6, this is the only interesting chart to study. We have

$$\frac{f(v,\eta)}{v^2\eta^2} := \frac{1}{u^2\eta^2}\left(v\eta + 1 + v^3\eta + v^4 - 2v^2\eta - 2v^2\right).$$

Then it is easy to see that $f(v,0)$ has the two double roots ± 1. Additionally $(v-1)^2$ also divides $f(v,\eta)$, whereas $f(-1,\eta) \neq 0$. Now Corollary 3.3.2.8 implies that the regular parts T_α are isomorphic for all $\alpha \in \Gamma_2$. Additionally $T_{\alpha_2} \neq 0$. Next consider the regular part T_{α_1} with respect to (ρ, α_1). This means, we have to consider

$$\Psi_\eta\left(\mathscr{E}^{\varphi(u)+\psi\left(\frac{\rho(\eta)}{u}\right)-\alpha_1(\eta)} \otimes (R \boxtimes S)\right).$$

It is

$$\varphi(u) + \psi\left(\frac{\rho(\eta)}{u}\right) - \alpha_1(\eta) = \frac{1}{u^2\eta^4}\left(u\eta^4 + \eta^4 + u^3\eta^2 + u^4 + 2u^2\eta^2\right).$$

Blowing up (u,η) with the help of the map e_0 defined above gives $\frac{f(v,\eta)}{v^2\eta^2}$, where

$$f(v,\eta) := v\eta + 1 + v^3\eta + v^4 + 2v^2.$$

Now the double roots of $f(v,0)$ are $\pm i$. Moreover $(v \pm i)^2$ divides $f(v,\eta)$. Note that because of $\alpha_1(-u) = \alpha_1(u)$, Γ_1 only contains $\alpha_1(u)$ (twice). So we do not need to use Corollary 3.3.2.8. It is $T_{\alpha_1} \neq 0$.
Now we can apply Theorem 3.3.2.10 and get the decomposition

$$\left(\mathrm{El}\left(\mathrm{Id}, \varphi(x), R\right) *_0 \mathrm{El}\left(\mathrm{Id}, \psi(y), S\right)\right)_0^{\wedge,\mathrm{loc.}} \cong \bigoplus_{k=1}^{2} \mathrm{El}\left(\rho, \alpha_k, T_{\alpha_k}\right).$$

3.3.2.14 Example: Let

$$\varphi(x) = \frac{1}{x} + \frac{1}{3x^3} \text{ and } \psi(y) = \frac{1}{2y^2}.$$

Hence $n = 3$, $m = 2$. With Proposition 3.3.1.2, $C := \{(x,y) | x\varphi'(x) - y\psi'(y) = 0\}$. Applying the direct image under $\gamma : (x,y) \mapsto (x, xy) =: (u,z)$, our equation changes to

$$\frac{u^2}{z^2} - \frac{1}{u} - \frac{1}{u^3} = 0 \text{ resp. } P(u,z) := u^5 - \left(1 + u^2\right) z^2 = 0.$$

The lowest u-exponent of the coefficient of z^0 is 5, the lowest u-exponent of the coefficient z^2 is 0 (the coefficient of z is equal to zero, therefore, we leave it out).

So the Newton polygon of P only has the vertices $(0,5)$ and $(2,0)$, i.e., it just has the slope $\frac{5}{2}$ different from 0 and ∞. Now we consider the polynomial

$$P\left(u, c_1 u^{\frac{5}{2}} + \tilde{z}_1\right) = u^5 - c_1^2 u^5 - c_1^2 u^7 - \left(2c_1 u^{\frac{5}{2}} + 2c_1 u^{\frac{9}{2}}\right)\tilde{z}_1 - \left(1+u^2\right)\tilde{z}_1^2.$$

The term $u^5 - c_1^2 u^5$ has to vanish. Hence $c_1 = \pm 1$. Continuing with this method gives the Puiseux series

$$z_\pm(u) = \pm u^{\frac{5}{2}} \mp \frac{1}{2}u^{\frac{9}{2}} \pm \frac{3}{8}u^{\frac{13}{2}} \mp \frac{5}{16}u^{\frac{17}{2}} \pm \frac{35}{128}u^{\frac{21}{2}} \mp \frac{63}{256}u^{\frac{25}{2}} \pm \ldots$$

Note that it is enough to compute the first $m^*n = 6$ terms of the parametrizations. Thus the representations of $z_\pm$ given above are sufficient (in the sense that we are able to compute the corresponding meromorphic parts $\alpha_\pm$). Next we apply Remark 3.3.2.2: Here it is $\gcd(n,m) = 1$. Hence $P(u,z)$ only has one irreducible component. So it is sufficient to continue with $z_+(u)$. Then the pair

$$\left(t^2, \tilde{\rho}_+(t)\right) := \left(t^2, t^5 - \frac{1}{2}t^9 + \frac{3}{8}t^{13} - \frac{5}{16}t^{17} + \frac{35}{128}t^{21} - \frac{63}{256}t^{25} + \ldots\right)$$

solves the equation $P(u,z) = 0$. The next step is the computation of $\alpha_+(t)$. Let $\lambda_+(t) \in t\mathbb{C}[\![t]\!]$ with $\lambda'(0) \neq 0$ and $\rho_+(\lambda_+(t)) = t^5 =: \rho(t)$. So we have

$$\alpha_+(t) := \varphi\left(\lambda_+^2\right) + \psi\left(\frac{t^5}{\lambda_+^2}\right) = \frac{1}{\lambda_+^2} + \frac{1}{3}\frac{1}{\lambda_+^6} + \frac{1}{2}\frac{\lambda_+^4}{t^{10}} \bmod \mathbb{C}[\![t]\!].$$

With Theorem 3.3.2.4,

$$\left(\rho^+\left(\mathrm{El}\left([x \mapsto x], \varphi(x), R\right) *_0 \mathrm{El}\left([y \mapsto y], \psi(y), S\right)\right)\right)_0^{\wedge,\mathrm{loc.}} \cong \bigoplus_{\alpha\in\Gamma} \mathscr{E}^{\alpha} \otimes T_\alpha,$$

where $\Gamma = \{\alpha_+(\xi t)\,|\,\xi^5 = 1\}$ is already decomposed in disjoint subsets in the sense of Theorem 3.3.2.4. Due to Corollary 3.3.2.8, the corresponding regular parts T_α are isomorphic for all $\alpha \in \Gamma$. So Theorem 3.3.2.10 implies

$$\left(\mathrm{El}\left(\mathrm{Id}, \varphi(x), R\right) *_0 \mathrm{El}\left(\mathrm{Id}, \psi(y), S\right)\right)_0^{\wedge,\mathrm{loc.}} \cong \mathrm{El}\left(\rho, \alpha_+, T_{\alpha_+}\right)$$

with a regular connection T_{α_+}.

4. The additive convolution of elementary mer. connections

In this chapter, we want to give a formula for the local formal description of the additive convolution of elementary formal meromorphic connections:

4.1 Definition: Let $\mathrm{El}(\rho,\varphi,R)$, $\mathrm{El}(\eta,\psi,S)$ be elementary formal meromorphic connections over $\mathbb{P}^1$ with $\rho \in x\mathbb{C}[\![x]\!], \varphi \in \mathbb{C}(\!(x)\!), \eta \in y\mathbb{C}[\![y]\!], \psi \in \mathbb{C}(\!(y)\!)$. Then, considering Remark 3.2 (2), the *additive convolution* $\mathrm{El}(\rho,\varphi,R) *_{+_i} \mathrm{El}(\eta,\psi,S)$ of degree $i \in \mathbb{Z}$ is defined as

$$\mathscr{H}^i\pi_+\gamma_+\left(\left((\rho_+\mathscr{E}^\varphi)\otimes\tilde{R}\right)\boxtimes\left((\eta_+\mathscr{E}^\psi)\otimes\tilde{S}\right)\right),$$

where $\gamma : \mathbb{C}^2 \to \mathbb{C}^2, (x,y) \mapsto (x, x+y) =: (u,z)$ and $\pi : \mathbb{C}^2 \to \mathbb{C}, (u,z) \mapsto z$.

As in Chapter 3, we are satisfied with computing the convolution of degree 0 as it is the only interesting cohomology module. Additionally we set $\rho = \mathrm{Id}$, $\eta = \mathrm{Id}$ again. This means, we want to write

$$\left(\mathscr{H}^0\pi_+\gamma_+\left((\mathscr{E}^\varphi\otimes R)\boxtimes(\mathscr{E}^\psi\otimes S)\right)\right)^\wedge_0$$

in the form

$$\bigoplus_k \mathrm{El}\left(\rho_k,\alpha_k,T_{\alpha_k}\right).$$

To do so, we use the know-how from Chapter 3. Section 3.3 tells us that the computation of the formal irregular part of

$$\left(\mathrm{El}(\mathrm{Id},\varphi,R) *_{+_0} \mathrm{El}(\mathrm{Id},\psi,S)\right)^\wedge_0,$$

i.e., the formalization of the module $\mathscr{H}^0\pi_+\gamma_+\left(\mathscr{E}^{\varphi(x)+\psi(y)}\otimes(R\boxtimes S)\right) = \mathscr{H}^0 f_+\left(\mathscr{E}^g\otimes T\right)$ with $f := \pi\circ\gamma$, $g := \varphi(x)+\psi(y)$, $T := R\boxtimes S$, at the origin of the disc Δ, can be reduced to a simpler problem: Consider the diagram

$$U \xrightarrow{(f,g)} \mathbb{C}\times\mathbb{C} \overset{i}{\hookrightarrow} \mathbb{P}^1\times\mathbb{P}^1 \overset{i_0}{\hookleftarrow} \Delta\times\mathbb{P}^1.$$

So the new problem is the parametrization of the singular locus of $\mathscr{H}^0 i_+(f,g)_+T$ in the neighbourhood of $(0,\infty)$. Note that the proof of Lemma 3.3.1.1 is completely analogous in the case of additive convolution, hence f and g are algebraically independent. So we are allowed to apply Proposition 3.3.1.2 which gives that the singular locus of the module $\mathscr{H}^0 i_+(f,g)_+T$ is the set

$$\begin{aligned}&\operatorname{Supp}\left(\mathscr{H}^0 i_+(f,g)_+T\right) \setminus \left(\left(\{0\} \times \mathbb{P}^1\right) \cup \left(\Delta \times \{\infty\}\right)\right)\\ =\;&\overline{(f,g)(C)} \setminus \left(\left(\{0\} \times \mathbb{P}^1\right) \cup \left(\Delta \times \{\infty\}\right)\right)\end{aligned}$$

with

$$C = \left\{ (x,y) \,\middle|\, \det \begin{pmatrix} \frac{\partial f(x,y)}{\partial x} & \frac{\partial g(x,y)}{\partial x} \\ \frac{\partial f(x,y)}{\partial y} & \frac{\partial g(x,y)}{\partial y} \end{pmatrix} = 0 \right\}.$$

This means, for $f(x,y) = x + y, g(x,y) = \varphi(x) + \psi(y)$,

$$C = \left\{(x,y) | \varphi'(x) = \psi'(y)\right\}.$$

Next we set

$$\varphi(x) := \sum_{i=1}^{n} a_i \frac{1}{x^i} \text{ and } \psi(y) := \sum_{j=1}^{m} b_j \frac{1}{y^j}$$

with $a_n \neq 0$ and $b_m \neq 0$. So C is the set of points $(x,y) \in \mathbb{C}^2$ with

$$\sum_{i=1}^{n} i a_i x^{-i-1} - \sum_{j=1}^{m} j b_j y^{-j-1} = 0. \tag{4.1}$$

Remember that in Chapter 3, we applied γ_+ with $\gamma : (x,y) \mapsto (x, xy) =: (u,z)$ first and then parameterized the resulting equation. Here, for the sake of simplicity, we will parameterize equation (4.1) first and apply γ_+ with $\gamma : (x,y) \mapsto (x, x+y) =: (u,z)$ afterwards.

4.2 Lemma: The equation

$$\sum_{i=1}^{n} i a_i x^{-i-1} - \sum_{j=1}^{m} j b_j y^{-j-1} = 0$$

has exactly $m+1$ parametrizations of the form

$$(x, y_k(x)) = \left(x, \sum_{p=1}^{\infty} c_{p,k} x^{\frac{n+1+\gcd(m+1,n+1)(p-1)}{m+1}} \right), \quad k = 1, ..., m+1.$$

Proof: First we multiply the equation

$$\sum_{i=1}^{n} ia_i x^{-i-1} - \sum_{j=1}^{m} jb_j y^{-j-1} = 0$$

by $x^{n+1}y^{m+1}$. Then it is equivalent to the equation $P(x, y) = 0$ with

$$P(x,y) = mb_m x^{n+1} + (m-1)b_{m-1}x^{n+1}y + ... + b_1 x^{n+1}y^{m-1} + \left(\sum_{i=1}^{n} -ia_i x^{n-i}\right) y^{m+1}.$$

As we want to find parametrizations $(x, y(x))$, we consider $P(x, y)$ as a polynomial in y with coefficients in $\mathbb{C}[x]$. Now, just as in the proof of Lemma 3.3.2.1, we have to find the lowest x-exponents in the coefficients of y^j - these are $n+1$ for all $j \in \{0, ..., m-1\}$ and 0 for $j = m+1$ (the coefficient of y^m equals zero). Connecting the points $(j, n+1)$, $j \in \{0, ..., m-1\}$, and $(m+1, 0)$ does not give a convex graph. Hence the Newton polygon has the slope $\frac{n+1}{m+1}$ (as in the proof of Lemma 3.3.2.1, all slopes are negative, but we only talk about their absolute values). Hence we set $y(x) = c_1 x^{(n+1)/(m+1)} + \tilde{y}_1$ with a new variable $\tilde{y}_1$. Next we consider the function $P\left(x, c_1 x^{(n+1)/(m+1)} + \tilde{y}_1\right)$ and choose c_1 such that the term with the lowest x-exponent in the coefficient of $\tilde{y}_1^0$ vanishes. This gives $c_{1,k} = \zeta_{m+1}^k \sqrt[m+1]{\frac{mb_m}{na_n}}$, $k \in \{1, ..., m+1\}$ (note that $a_n, b_m \neq 0$). So we have found $m+1$ starting terms which lead to parametrizations $y_k(x)$ for $k = 1, ..., m+1$ due to [14]. As $P(x, y)$ is a polynomial of degree $m+1$ in y, there can be no more parametrizations. Now we want to show that during the further procedure all Newton polygons have exactly two slopes - the used slope $\frac{n+1}{m+1}$ and a new slope strictly greater than $\frac{n+1}{m+1}$ which is an integer multiple of $\frac{\gcd(n+1,m+1)}{m+1}$. We will prove this by induction over the number q of executed steps of the algorithm. For the sake of simplicity, we leave out the index $k \in \{1, ..., m+1\}$ (because we already know that there are exactly $m+1$ parametrizations) and just use the index q. Before starting the induction, we need some notation: We denote by $P_1(x, \tilde{y}_1) := P\left(x, \zeta_{m+1} \sqrt[m+1]{\frac{mb_m}{na_n}} x^{(n+1)/(m+1)} + \tilde{y}_1\right) \in \left(\mathbb{C}\left[x^{(n+1)/(m+1)}\right]\right)[\tilde{y}_1]$ the function which we receive after the construction of the first Newton polygon at the beginning of the proof. In analogy, we denote by $P_q(x, \tilde{y}_q) := P_{q-1}\left(x, c_q x^{\gamma_q} + \tilde{y}_q\right)$ the function which we receive after the construction of the q-th Newton polygon.

We want to show that for all $q \in \mathbb{N}$, the function $P_q(x, \tilde{y}_q)$ has exactly the two slopes $\frac{n+1}{m+1}$ and $\gamma_{q+1} > \gamma_q$, where γ_{q+1} is an integer multiple of $\frac{\gcd(n+1,m+1)}{m+1}$. To do this, we show the following three points:

1) For all $q \in \mathbb{N}$, the lowest x-exponents in the coefficients of $\tilde{y}_q^j$ in the function $P_q(x, \tilde{y}_q)$ are $\frac{n+1}{m+1}(m+1-j)$, $j \in \{1, ..., m+1\}$.

2) For all $q \in \mathbb{N}$, all x-exponents in the coefficient of $\tilde{y}_q^0$ in the function $P_q(x, \tilde{y}_q)$ are integer multiples of $\frac{\gcd(n+1,m+1)}{m+1}$ greater than $\gamma_q + m\frac{n+1}{m+1}$ (then we have $\gamma_{q+1} > \gamma_q$).

3) For all $q \in \mathbb{N}$, the defining equation for the coefficient c_{q+1} is linear in c_{q+1}, so c_{q+1} is uniquely determined.

For the induction basis, consider $P_1(x, \tilde{y}_1) = P\left(x, \zeta_{m+1} \sqrt[m+1]{\frac{mb_m}{na_n}} x^{(n+1)/(m+1)} + \tilde{y}_1\right)$. Computing $P_1(x, \tilde{y}_1)$, we get a function in which the coefficient of $\tilde{y}_1^j$ for $j = 1, ..., m+1$ consists of a sum of terms with x-exponents

i) $n + 1 + l\frac{n+1}{m+1}$ and $2n - i + 1$ for $l \in \{1, ..., m-1\}, i \in \{1, ..., n-1\}$, if $j = 0$, and

ii) $n + 1 + (l - j)\frac{n+1}{m+1}$ and $n - i + \frac{n+1}{m+1}(m + 1 - j)$ for $l \in \{j, ..., m-1\}$, $i \in \{1, ..., n\}$, if $j \in \{1, ..., m+1\}$.

Note that, due to the construction of c_1, for $j = 0$ the terms with x-exponent $n + 1$ vanish, so the lowest x-exponent in the coefficient of $\tilde{y}_1^0$ is t_1 which is an integer of $\frac{\gcd(n+1,m+1)}{m+1}$ and greater than $n + 1$. Moreover, as $P(x, y)$ only contained integer x-exponents, condition 2) is clearly true. The lowest x-exponent in the coefficient of $\tilde{y}_1^{m+1}$ can directly seen to be 0. It remains to find the lowest x-exponent in every coefficient of the $\tilde{y}_1^j$ for $j = 1, ..., m$. They are given by $(m+1-j)\frac{n+1}{m+1}$ - these are exactly the x-terms which contain $a_n \neq 0$ in the coefficient (besides other non-zero factors), hence they are not zero. Thus also condition 1) holds. So the Newton polygon of $P_1(x, \tilde{y}_1)$ has the two slopes $\frac{n+1}{m+1}$ and $\gamma_2 := t_1 - m\frac{n+1}{m+1} > \frac{n+1}{m+1} = \gamma_1$ (as $t_1 > n + 1$). Now let

$$\bar{P}_1(x, \tilde{y}_1) := x^{t_1} + x^{m\frac{n+1}{m+1}}\tilde{y}_1 + ... + x^{\frac{n+1}{m+1}}\tilde{y}_1^m + \tilde{y}_1^{m+1}$$

be the auxiliary function which contains only the terms that define the Newton polygon and no constants. Set $\tilde{y}_1 := c_2 x^{\gamma_2} + \tilde{y}_2$. The function $\bar{P}_2(x, \tilde{y}_2) := \bar{P}_1(x, c_2 x^{\gamma_2} + \tilde{y}_2)$ suffices to compute c_2 because c_2 only has to fulfill the property that the term with the lowest x-exponent t_1 in the coefficient of $\tilde{y}_2^0$ in $P_2(x, \tilde{y}_2)$ vanishes. In $\bar{P}_2(x, \tilde{y}_2)$, the coefficient of $\tilde{y}_2^0$ consists of terms with the x-exponents t_1 and $(m+1-j)\frac{n+1}{m+1} + j\gamma_2$ for $j = 1, ..., m+1$. But

$$\begin{aligned}(m+1-j)\frac{n+1}{m+1} + j\gamma_2 &= (m+1-j)\frac{n+1}{m+1} + j\left(t_1 - m\frac{n+1}{m+1}\right)\\ = n + 1 - j\frac{n+1}{m+1} + jt_1 - jm\frac{n+1}{m+1} &= n + 1 - j(n+1) + jt_1 > t_1 \text{ if } j > 1.\end{aligned}$$

This means, the defining equation for c_2 is linear in c_2, i.e., an equation of the form $a + c_2 b = 0$ for $a, b \in \mathbb{C}$. Hence c_2 is uniquely determined. This proves condition 3).

For the induction step, we want to show that the function $P_q(x, \tilde{y}_q)$ fulfills the three conditions. By definition, $P_q(x, \tilde{y}_q) = P_{q-1}(x, c_q x^{\gamma_q} + \tilde{y}_q)$. By the induction hypothesis, $P_{q-1}(x, \tilde{y}_{q-1})$ has the three desired properties. As $P_q(x, \tilde{y}_q)$ is given, we know $c_q x^{\gamma_q}$. So

$$\begin{aligned} P_q(x, \tilde{y}_q) =& P_{q-1}(x, c_q x^{\gamma_q} + \tilde{y}_q) = f_{q-1}\left(x^{\frac{\gcd(n+1,m+1)}{m+1}}\right) + \left(d_1 x^{m\frac{n+1}{m+1}} + \dots\right)(c_q x^{\gamma_q} + \tilde{y}_q) \\ &+ \dots + \left(d_m x^{\frac{n+1}{m+1}} + \dots\right)(c_q x^{\gamma_q} + \tilde{y}_q)^m - \left(\sum_{i=1}^{n} i a_i x^{n-i}\right)(c_q x^{\gamma_q} + \tilde{y}_q)^{m+1} \end{aligned}$$

with a polynomial $f_{q-1}(s) \in \mathbb{C}[s]$ such that $f_{q-1}\left(x^{\frac{\gcd(m+1,n+1)}{m+1}}\right)$ has property 2) and $d_j \in \mathbb{C}^\times$ for all $j \in \{1, \dots, m\}$. Considering the coefficients of $\tilde{y}_q^j$ for $j \in \{1, \dots, m+1\}$, the lowest x-exponents are $(m+1-j)\frac{n+1}{m+1}$ again, thus $P_q(x, \tilde{y}_q)$ has property 1). The coefficient of $\tilde{y}_q^0$ contains $f\left(x^{\frac{\gcd(m+1,n+1)}{m+1}}\right)$ (which has x-exponents that are greater than $\gamma_{q-1} + m\frac{n+1}{m+1}$ due to property 1)) and new terms with x-exponents

$$\begin{aligned} &(m+1-j)\frac{n+1}{m+1} + j\gamma_q > \gamma_{q-1} + m\frac{n+1}{m+1} \\ \Leftrightarrow\, & n+1-j\frac{n+1}{m+1} + j\gamma_q > \gamma_{q-1} + m\frac{n+1}{m+1} \\ \Leftrightarrow\, & j\left(\gamma_q - \frac{n+1}{m+1}\right) > \gamma_{q-1} + m\frac{n+1}{m+1} - n - 1 \\ \Leftrightarrow\, & j\left(\gamma_q - \frac{n+1}{m+1}\right) > \gamma_{q-1} - \frac{n+1}{m+1} \quad \text{(this equation holds as } \gamma_q > \gamma_{q-1}) \end{aligned}$$

(and greater ones), hence $P_q(x, \tilde{y}_q)$ has property 2) because by the induction hypothesis, γ_q is an integer multiple of $\frac{\gcd(n+1,m+1)}{m+1}$ and because the term, which has the x-exponent $\gamma_{q-1} + m\frac{n+1}{m+1}$, vanishes due to the choice of c_q. This gives the fact that the Newton polygon has the two slopes $\frac{n+1}{m+1}$ and $\gamma_{q+1} > \gamma_q$ with $\gamma_{q+1} \in \frac{\gcd(n+1,m+1)}{m+1}\mathbb{N}$. Finally property 3) holds which can be shown as in the induction basis. By induction, we get the desired parametrizations

$$(x, y_k(x)) = \left(x, \sum_{p=1}^{\infty} c_{p,k} x^{\frac{n+1+\gcd(m+1,n+1)(p-1)}{m+1}}\right), \quad k = 1, \dots, m+1.$$

□

Now, after applying γ_+ with $\gamma : (x, y) \mapsto (x, x + y) =: (u, z)$, the set

$$C = \left\{ (x, y) \middle| \sum_{i=1}^{n} i a_i x^{-i-1} - \sum_{j=1}^{m} j b_j y^{-j-1} = 0 \right\}$$

changes to

$$C' := \left\{ (u, z) \middle| \sum_{i=1}^{n} i a_i u^{-i-1} - \sum_{j=1}^{m} j b_j (z - u)^{-j-1} = 0 \right\}.$$

In order to find the Puiseux parametrizations of this set, we use the construction from Lemma 4.2. If

$$(x, y(x)) = \left(x, \sum_{p=1}^{\infty} c_p x^{\frac{n+1+\gcd(m+1,n+1)(p-1)}{m+1}} \right)$$

is a parametrization of C, then the application of γ_+ with $\gamma : (x, y) \mapsto (x, x + y) = (u, z)$ changes this parametrization to

$$(u, z(u)) = \left(u, u + \sum_{p=1}^{\infty} c_p u^{\frac{n+1+\gcd(m+1,n+1)(p-1)}{m+1}} \right).$$

But here, a specific problem can occur. In the case $m = n$, the parametrization simply has the form
Now it is easy to see that the order of the power series $z(u)$ increases if $c_1 = -1$. This means, we have to distinguish between the three cases

1. $m < n$,

2. $m > n$ and

3. $m = n$.

Note that, as we do not compute the regular parts T_{α_k} in the decomposition of the convolution so precisely, we can assume $m \leq n$ without loss of generality because of

$$\mathscr{E}^{\varphi} *_{+0} \mathscr{E}^{\psi} = \mathscr{H}^0 \pi_+ \gamma_+ \left(\mathscr{E}^{\varphi} \boxtimes \mathscr{E}^{\psi} \right) \cong \mathscr{H}^0 \pi_+ \gamma_+ \left(\mathscr{E}^{\psi} \boxtimes \mathscr{E}^{\varphi} \right) = \mathscr{E}^{\psi} *_{+0} \mathscr{E}^{\varphi}.$$

So in Section 4.1, we will study the case $m < n$. Section 4.2 will treat the case $m = n$.

4.1. The case $m < n$

Corollary 4.1.1: The set C' (see page 53) has exactly $m+1$ parametrizations of the form

$$(u, z_k(u)) = \left(u, u + \sum_{p=1}^{\infty} c_{p,k} u^{\frac{n+1+\gcd(n+1,m+1)(p-1)}{m+1}}\right), \quad k \in \{1, ..., m+1\},$$

with $c_{1,k} = 1$ for all k.

Proof: It follows from Lemma 4.2 that all parametrizations of $P(u,z) = 0$ have the form $(u, z_k(u)) = \left(u, u + \sum_{p=1}^{\infty} c_{p,k} u^{\frac{n+1+\gcd(n+1,m+1)(p-1)}{m+1}}\right)$ with $c_{1,k} = \zeta_{m+1} \sqrt[m+1]{\frac{mb_m}{na_n}}$. All the other $c_{p,k}$ are uniquely determined for each k. Hence there are exactly $m+1$ parametrizations of the desired form. □

Next we want to use the substitution $\beta : t \mapsto t^q = u$ with $q = \frac{m+1}{\gcd(m+1,n+1)}$ such that the z_k have integer exponents. Hence the set

$$C' := \left\{ (u,z) \middle| \sum_{i=1}^{n} i a_i u^{-i-1} - \sum_{j=1}^{m} j b_j (z-u)^{-j-1} = 0 \right\}.$$

is parameterized by the pairs

$$(t^q, \tilde{\rho}_k(t)), \ k = 1, ..., m+1,$$

with $\tilde{\rho}_k(t) := z_k(t^q)$. In order to give the singular locus, let $P_1, ..., P_r$ be the pairwise distinct irreducible components of the decomposition of the polynomial $P(u,z)$ over $\mathbb{C}[\![u]\!][z]$ and $\tilde{\rho}_k(t)$ be a parametrization of P_k for all $k = 1, ..., r$. Clearly $r \leq m+1$. For a suitable renumbering of the parametrizations, the singular locus is the set

$$\{(\tilde{\rho}_k(t), \varphi(t^q) + \psi(\tilde{\rho}_k(t) - t^q)) | \, k = 1, ..., r\}.$$

To complete the parametrization, we have to compute the series

$$\varphi(t^q) + \psi(\tilde{\rho}_k(t) - t^q) = \sum_{i=1}^{n} a_i \frac{1}{t^{qi}} + \sum_{j=1}^{m} b_j \frac{1}{(\tilde{\rho}_k(t) - t^q)^j}, \quad k \in \{1, ..., r\}.$$

Note that $\varphi(t^q)$ has pole order qn, and $\psi(\tilde{\rho}_k(t) - t^q)$ has pole order $(n+1)\frac{qm}{m+1}$. We have $qn > (n+1)\frac{qm}{m+1}$ because $n > m$. Additionally this shows that a_n is the coefficient of the term $\frac{1}{t^{qn}}$. So, using polynomial division, the results are the Laurent series

$$\underbrace{\sum_{l=1}^{qn} d_{l,k} t^{-l}}_{=:\tilde{\alpha}_k(t) \text{ (meromorphic part)}} + \underbrace{\sum_{l=0}^{\infty} \tilde{d}_{l,k} t^{l}}_{=:\tilde{\delta}_k(t) \text{ (holomorphic part)}} , k \in \{1, ..., r\},$$

with $d_{qn,k} = a_n \ \forall k$. Next let $\lambda_k(t) \in t\mathbb{C}[\![t]\!]$ with $\lambda_k'(0) \neq 0$ and $\rho(t) := (\tilde{\rho}_k \circ \lambda_k)(t) = t^q$ be given for all k. This changes $\tilde{\alpha}_k + \tilde{\delta}_k$ to functions $\left(\left(\tilde{\alpha}_k + \tilde{\delta}_k\right) \circ \lambda_k\right)(t)$ which we also decompose into a meromorphic part α_k (also of pole order qn) and a holomorphic part δ_k. Just as in Chapter 3, we have to cancel the redundant parametrizations, i.e., the parametrizations for which there exists an q-th root of unity ζ such that we have an equality of the form $\alpha_k(\zeta t) + \delta_k(\zeta t) = \alpha_l(t) + \delta_l(t)$ for $k, l \in \{1, ..., r\}$ with $k \neq l$. Assume that after this procedure, we have $r^* \leq r$ parametrizations left. Then for a suitable renumbering of the parametrizations, we receive a minimal representation

$$\mathscr{S} := \{(t^q, \alpha_k(t) + \delta_k(t)) | k = 1, ..., r^*\}$$

of the singular locus. Again it is enough to consider the meromorphic parts $\alpha_k(t)$ (see Def. 3.1), i.e., the pairs

$$(\rho(t), \alpha_k(t)), \ k \in \{1, ..., r^*\}.$$

Now we can apply Theorem 3.3.2.4 and get a decomposition

$$\left(\rho^+ \left(\mathrm{El}(\mathrm{Id}, \varphi, R) *_{+0} \mathrm{El}(\mathrm{Id}, \psi, S)\right)\right)_0^\wedge \cong \bigoplus_{\alpha \in \Gamma} \mathscr{E}^\alpha \otimes T_\alpha$$

with

$$\Gamma = \{\alpha_k(\xi t) | k \in \{1, ..., r^*\}, \xi^q = 1\}.$$

Γ contains at most qr^* distinct elements. As the elements $\alpha_k(\xi t)$ with $k \in \{1, ..., r^*\}$ and $\xi^q = 1$ do not have to be pairwise non-identical, Γ can be decomposed in h disjoint subsets, $\Gamma = \biguplus_{k=1}^h \Gamma_k$, with

$$\Gamma_k := \{\alpha \in \Gamma | \exists \xi \in \mathbb{C} : \xi^q = 1 \wedge \alpha(\xi t) = \alpha_k(t)\}.$$

So clearly $h \leq r^* \leq r \leq m + 1$. Next we want to show a decomposition of the form

$$\left(\mathrm{El}(\mathrm{Id}, \varphi(x), R) *_{+0} \mathrm{El}(\mathrm{Id}, \psi(y), S)\right)_0^\wedge \cong \bigoplus_{k=1}^{h} \mathrm{El}(\rho, \alpha_k, T_k).$$

We have shown in Section 3.3.2 that the problem reduces to the computation of

$$\Psi_\eta\left(\mathscr{E}^{\varphi(u)+\psi(\rho(\eta)-u)-\alpha_k(\eta)}\otimes(R\boxtimes S)\right).$$

But in analogy to the proof of Lemma 3.3.2.9, we can equivalently compute the module

$$\Psi_\eta\left(\mathscr{E}^{\varphi(u)+\psi(\tilde{\rho}_k(\eta)-u)-\tilde{\alpha}_k(\eta)}\otimes(R\boxtimes S)\right)$$

with $\tilde{\rho}_k, \tilde{\alpha}_k$ defined on page 55. Remember the notation

$$\varphi(u)=\sum_{i=1}^{n}a_i\frac{1}{u^i},\ \psi(\tilde{\rho}_k(\eta)-u)=\sum_{j=1}^{m}b_j\frac{1}{(\tilde{\rho}_k(\eta)-u)^j},\ \tilde{\alpha}_k(\eta)=\sum_{l=1}^{qn}d_{l,k}\frac{1}{\eta^l}$$

with $d_{qn,k}=a_n$ for all k. Then we have

$$\begin{aligned}&\varphi(u)+\psi(\tilde{\rho}_k(\eta)-u)-\tilde{\alpha}_k(\eta)\\ =&\frac{1}{u^n\eta^{qn}}\left(\eta^{nq}a(u)+\eta^{qn}u^n\sum_{j=1}^{m}b_j\frac{1}{(\tilde{\rho}_k(\eta)-u)^j}-u^nd_k(\eta)\right),\end{aligned}$$

where $a(u):=\sum_{i=1}^{n}a_iu^{n-i}$ and $d_k(\eta):=\sum_{l=1}^{qn}d_{l,k}\eta^{qn-l}$ are holomorphic functions. Now the singularity at $(u,\eta)=(0,0)$ has to be solved. We use the setting of Lemma 3.3.2.6.

4.1.2 Lemma: For solving the singularity, one needs a composition of q blowups. This gives a sequence of charts $e_i:(v,\eta)\mapsto(v\eta^{q-i},v^i\eta)$, $i=0,...,q$. Then for all $i>0$, the corresponding nearby cycles module is trivial.

Proof: The composition of blowups gives a resolution $\nu=(\nu_1,\nu_2):\mathbb{X}\to\mathbb{C}^2$. Due to Lemma 3.3.2.6(1),

$$\Psi_\eta\left(\mathscr{E}^{\varphi(u)+\psi(\tilde{\rho}_k(\eta)-u)-\tilde{\alpha}_k(\eta)}\otimes T\right)\cong\tilde{\nu}_+\Psi_{\eta\circ\nu}\left(\iota_+\nu^+\left(\mathscr{E}^{\varphi(u)+\psi(\tilde{\rho}_k(\eta)-u)-\tilde{\alpha}_k(\eta)}\otimes T\right)\right)$$

with $\tilde{\nu}:\mathbb{X}\times\{0\}\to\mathbb{C}^2\times\{0\}$ and $\iota=(\mathrm{Id}_{\mathbb{X}},\nu_2)$. Together with the resolution ν, we get the sequence of charts $e_i:(v,\eta)\mapsto(v\eta^{q-i},v^i\eta)$, $i=0,...,q$. Let $i>0$. e_i^+ applied to the module $\mathscr{E}^{\varphi(u)+\psi(\tilde{\rho}_k(\eta)-u)-\tilde{\alpha}_k(\eta)}$ gives

$$\frac{1}{v^{iqn}\eta^{qn}\left(\tilde{\rho}_k\left(v^i\eta\right)/v\eta^{q-i}-1\right)^m}f_{k,i}(v,\eta)$$

with

$$
\begin{aligned}
f_{k,i}(v,\eta) =& v^{n(iq-1)}\eta^{in} a\left(v\eta^{q-i}\right)\left(\tilde{\rho}_k\left(v^i\eta\right)/v\eta^{q-i}-1\right)^m \\
&+\sum_{j=1}^{m} b_j v^{iqn-j}\eta^{qn-jq+ij}\left(\tilde{\rho}_k\left(v^i\eta\right)/v\eta^{q-i}-1\right)^{m-j} \\
&-d_k\left(v^i\eta\right)\left(\tilde{\rho}_k\left(v^i\eta\right)/v\eta^{q-i}-1\right)^m .
\end{aligned}
$$

Note that $\tilde{\rho}_k\left(v^i\eta\right)/v\eta^{q-i}$ is holomorphic. By definition, the module $\Psi_{\eta\circ e_i}\left(e_i^+T\right)$ is supported on $\{v^i\eta=0\}$. Moreover it is even supported on $\{\eta=0\}$ (see the setting of Proposition 2.1). But at every point, we can change coordinates in order to get the situation in [7, Lemma 7.3]. Hence the Lemma follows with Proposition 2.1 because we have

$$
f_{k,i}(v,0) = -d_{qn,k} = -a_n \neq 0 \ \forall k=1,...,h, \ \forall i=1,...,q,
$$

i.e., $f_{k,i}(v,0)$ is an invertible constant. □

According to Lemma 4.1.2, we just have to consider the chart $e_0 : (v,\eta) \mapsto (v\eta^q,\eta)$. With $\sigma_k(u,\eta) := \varphi(u) + \psi(\tilde{\rho}_k(\eta)-u) - \tilde{\alpha}_k(\eta)$, this leads to the computation of the module $\Psi_\eta\left(\iota_+\left(\mathscr{E}^{\sigma_k(v\eta^q,\eta)}\otimes e_0^+T\right)\right)$ with

$$
\begin{aligned}
\sigma_k(v\eta^q,\eta) =& \frac{1}{v^n\eta^{2qn}}\left(\eta^{qn}a(v\eta^q)+\eta^{2qn}v^n\sum_{j=1}^{m} b_j\frac{1}{(\tilde{\rho}_k(\eta)-v\eta^q)^j}-v^n\eta^{qn}d_k(\eta)\right) \\
=& \frac{1}{v^n\eta^{qn}}\left(a(v\eta^q)+v^n\sum_{j=1}^{m} b_j\frac{\eta^{q(n-j)}}{\left(\frac{\tilde{\rho}_k(\eta)}{\eta^q}-v\right)^j}-v^n d_k(\eta)\right) \\
=& \frac{1}{\left(\frac{\tilde{\rho}_k(\eta)}{\eta^q}-v\right)^m v^n\eta^{qn}}\left(\left(\frac{\tilde{\rho}_k(\eta)}{\eta^q}-v\right)^m\left(a(v\eta^q)+v^n\sum_{j=1}^{m} b_j\frac{\eta^{q(n-j)}}{\left(\frac{\tilde{\rho}_k(\eta)}{\eta^q}-v\right)^j}-v^n d_k(\eta)\right)\right) \\
=:& \frac{f_k(v,\eta)}{\left(\frac{\tilde{\rho}_k(\eta)}{\eta^q}-v\right)^m v^n\eta^{qn}}.
\end{aligned}
$$

Note that this function has indeed no singularity at $(0,0)$ anymore because $f(0,0)=1$ as $\frac{\rho(\eta)}{\eta^q}|_{\eta=0}=1$. As ι is proper, we can equivalently compute

$$
\Psi_\eta\left(\mathscr{E}^{\sigma_k(v\eta^q,\eta)}\otimes e_0^+T\right).
$$

Alltogether the computation of

$$\Psi_\eta\left(\mathscr{E}^{\varphi(u)+\psi(\tilde{\rho}_k(\eta)-u)-\tilde{\alpha}_k(\eta)}\otimes T\right)$$

reduces to the computation of

$$\Psi_\eta\left(\mathscr{E}^{f_k(v,\eta)/\left(\frac{\tilde{\rho}_k(\eta)}{\eta^q}-v\right)^m v^n\eta^{qn}}\otimes e_0^+T\right).$$

Again this module is supported on $\eta=0$ and at most on the set $\{f_k(v,0)=0\}$.

4.1.3 Proposition: For all $k=1,...,h$, $\tilde{f}_k(v,\eta):=\left(\frac{\tilde{\rho}_k(\eta)}{\eta^q}-v\right)^{1-m}f_k(v,\eta)$ has one single double root at $v=1$. Moreover $(v-1)^2$ divides $\tilde{f}_k(v,\eta)$.

Proof: Because of $n>m$, all the b_j-terms in $f_k(v,0)$ vanish. We have

$$f_k(v,0)=(1-v)^m\left(a_n-v^n d_{qn,k}\right)=a_n(1-v)^{m+1}\prod_{i=1}^{n-1}\left(v-\zeta_n^i\right)$$

because of $d_{qn,k}=a_n\neq 0$ for all k. So all the roots $\zeta_n^i, i\in\{1,...,n-1\}$, are simple and the branches of $f_k(v,\eta)$ at the points $(\zeta_n^i,0)$ are smooth and transversal to $\eta=0$. It follows that in local coordinates $(v',\eta):=(v-\zeta_n^i,\eta)$ centered at the points $(\zeta_n^i,0)$, the function $\frac{f_k(v,\eta)}{\left(\frac{\tilde{\rho}_k(\eta)}{\eta^q}-v\right)^m v^n\eta^{qn}}$ can be written as v'/η^{qn} because in some neighbourhood of $(\zeta_n^i,0)$, the term $\left(\frac{\tilde{\rho}_k(\eta)}{\eta^q}-v\right)^m$ is a unit. Due to Lemma 3.2.3, $\Psi_\eta\left(\mathscr{E}^{v'/\eta^{qn}}\otimes e_0^+T\right)=0$. It remains to check the root $v=1$. Of course, it is a double root of $f_k(v,0)$ of order $m+1\geq 2$. This is a problem because we need a root exactly of order 2. So let us write

$$\begin{aligned}
&f_k(v,\eta)\\
&=\left(\frac{\tilde{\rho}_k(\eta)}{\eta^q}-v\right)^m\left(a(v\eta^q)+v^n\sum_{j=1}^m b_j\frac{\eta^{q(n-j)}}{\left(\frac{\tilde{\rho}_k(\eta)}{\eta^q}-v\right)^j}-v^n d_k(\eta)\right)\\
&=\left(\frac{\tilde{\rho}_k(\eta)}{\eta^q}-v\right)^{m-1}\left(\left(\frac{\tilde{\rho}_k(\eta)}{\eta^q}-v\right)\left(a(v\eta^q)+v^n\sum_{j=1}^m b_j\frac{\eta^{q(n-j)}}{\left(\frac{\tilde{\rho}_k(\eta)}{\eta^q}-v\right)^j}-v^n d_k(\eta)\right)\right)\\
&=:\left(\frac{\tilde{\rho}(\eta)}{\eta^q}-v\right)^{m-1}\tilde{f}_k(v,\eta).
\end{aligned}$$

This gives

$$\frac{f_k(v,\eta)}{\left(\frac{\tilde{\rho}_k(\eta)}{\eta^q}-v\right)^m v^n\eta^{qn}} = \frac{\tilde{f}_k(v,\eta)}{v^n\eta^{qn}\left(\frac{\tilde{\rho}_k(\eta)}{\eta^q}-v\right)},$$

where $\tilde{f}_k(v,0)$ also has the root $v=1$, but exactly of multiplicity 2.
Finally we have to check whether $(v-1)^2$ divides $\tilde{f}_k(v,\eta)$, i.e., whether

$$\tilde{f}_k(1,\eta) = \left.\frac{\partial \tilde{f}_k(v,\eta)}{\partial v}\right|_{v=1} = 0.$$

Remember that $\tilde{\alpha}_k$ can be written in the form

$$\tilde{\alpha}_k(\eta) = \varphi(\eta^q) + \psi(\tilde{\rho}_k(\eta)-\eta^q) \bmod \mathbb{C}[\![\eta]\!].$$

So, even in consideration of the holomorphic terms of $\varphi(\eta^q)+\psi(\tilde{\rho}_k(\eta)-\eta^q)$, we can write $\tilde{f}_k(v,\eta)$ in the form

$$\begin{aligned}\tilde{f}_k(v,\eta) =& \left(\frac{\tilde{\rho}_k(\eta)}{\eta^q}-v\right)\sum_{i=1}^{n} a_i\left((v\eta^q)^{n-i} - v^n\eta^{q(n-i)}\right)\\ &+\left(\frac{\tilde{\rho}_k(\eta)}{\eta^q}-v\right)\sum_{j=1}^{m} b_j v^n\eta^{q(n-j)}\left(\left(\frac{\tilde{\rho}_k(\eta)}{\eta^q}-v\right)^{-j} - \left(\frac{\tilde{\rho}_k(\eta)}{\eta^q}-1\right)^{-j}\right),\end{aligned}$$

Hence $\tilde{f}_k(1,\eta)=0$. It remains to check whether $\left.\frac{\partial \tilde{f}_k(v,\eta)}{\partial v}\right|_{v=1}=0$. For the sake of simplicity, we forget the first factor $\left(\frac{\tilde{\rho}_k(\eta)}{\eta^q}-v\right)$ in $\tilde{f}_k(v,\eta)$ because

$$\left.\frac{\partial \tilde{f}_k(v,\eta)}{\partial v}\right|_{v=1} = 0 \Leftrightarrow \left.\frac{\partial}{\partial v}\frac{\tilde{f}_k(v,\eta)}{\frac{\tilde{\rho}_k(\eta)}{\eta^q}-v}\right|_{v=1} = 0.$$

Thus

$$\begin{aligned}\frac{\partial}{\partial v}\frac{\tilde{f}_k(v,\eta)}{\frac{\tilde{\rho}_k(\eta)}{\eta^q}-v} =& \sum_{i=1}^{n} a_i\left((n-i)(v\eta^q)^{n-i-1}\eta^q - nv^{n-1}\eta^{q(n-i)}\right)\\ &+\sum_{j=1}^{m} b_j n v^{n-1}\eta^{q(n-j)}\left(\left(\frac{\tilde{\rho}_k(\eta)}{\eta^q}-v\right)^{-j} - \left(\frac{\tilde{\rho}_k(\eta)}{\eta^q}-1\right)^{-j}\right)\\ &+\sum_{j=1}^{m} j b_j v^n\eta^{q(n-j)}\left(\frac{\tilde{\rho}_k(\eta)}{\eta^q}-v\right)^{-j-1}.\end{aligned}$$

Setting $v = 1$ gives the function

$$\sum_{i=1}^{n} -ia_i(\eta^q)^{n-i} + \sum_{j=1}^{m} jb_j\eta^{q(n-j)} \left(\frac{\tilde{\rho}_k(\eta)}{\eta^q} - 1\right)^{-j-1}$$
$$= \frac{1}{\eta^{q(n+1)}} \left(\sum_{i=1}^{n} -ia_i(\eta^q)^{-n-1} + \sum_{j=1}^{m} jb_j \left(\tilde{\rho}_k(\eta) - \eta^q\right)^{-j-1}\right).$$

But the term in brackets is just the term

$$\sum_{i=1}^{n} -ia_i u^{-i-1} + \sum_{j=1}^{m} jb_j \left(z - u\right)^{-j-1}$$

which defines the set C' (see page 53) - and for this equation, the pair $(u, z) = (\eta^q, \tilde{\rho}_k(\eta))$ is a solution for all k. Hence

$$\left.\frac{\partial \tilde{f}_k(v, \eta)}{\partial v}\right|_{v=1} = 0.$$

Computing the second derivation of $\tilde{f}_k(v, \eta)$ with respect to v gives the fact that $(v-1)^2$ divides $\tilde{f}_k(v, \eta)$, but $(v-1)^3$ does not. □

Due to Lemma 4.1.3, in a sufficiently small neighbourhood of the point $(v, \eta) = (1, 0)$, the function

$$\frac{\tilde{f}_k(v, \eta)}{v^n \eta^{qn} \left(\frac{\tilde{\rho}_k(\eta)}{\eta^q} - v\right)}$$

can be written in the form

$$\frac{(v-1)^2 \cdot \text{unit}}{\eta^{qn + \frac{n+1}{\gcd(n+1, m+1)}} \cdot \tau_k(\eta)}$$

with $\tau_k(\eta) \in \mathbb{C}[\![\eta]\!]$ and $\tau_k(0) \neq 0$ because $\frac{\tilde{\rho}_k(\eta)}{\eta^q} - 1 \in \eta^{\frac{n+1}{\gcd(n+1, m+1)}} \mathbb{C}[\![\eta]\!]$ is a series of order $\frac{n+1}{\gcd(n+1, m+1)}$ (Corollary 4.1.1). But $\tau(\eta)$ is a unit. So equivalently, in a neighbourhood of $(v, \eta) = (1, 0)$, we can write

$$\frac{\tilde{f}_k(v, \eta)}{v^n \eta^{qn} \left(\frac{\tilde{\rho}_k(\eta)}{\eta^q} - v\right)} = \frac{(v-1)^2 \cdot \text{unit}}{\eta^{qn + \frac{n+1}{\gcd(n+1, m+1)}}}$$

Now Lemma 3.2.3 proves that all occuring regular parts T_α in the decomposition on page 55 are not equal to zero. Additionally they are all isomorphic as $\bar{v} = 1$ is the only value with $(v - \bar{v})^2 | \tilde{f}_k(v, \eta)$. Thus we can formulate the main result of this section.

4.1.4 Theorem: Let $\varphi(x) = \sum_{i=1}^{n} a_i x^{-i}$ resp. $\psi(y) = \sum_{j=1}^{m} b_j y^{-j}$ be meromorphic functions with pole order n resp. m (i.e. $a_n \neq 0$ resp. $b_m \neq 0$) and $m < n$. Let $d := \gcd(n+1, m+1)$, $q := \frac{m+1}{d}$. Then

(I) After applying the direct image γ_+ with $\gamma : (x,y) \mapsto (x, x+y) =: (u,z)$ to the set $C = \{(x,y) | \varphi'(x) = \psi'(y)\}$, its r pairwise distinct irreducible components have parametrizations of the form

$$\left(t^q, t^q + \sum_{p=1}^{\infty} c_{p,k} t^{\frac{n+1}{d}+p-1}\right), \quad k \in \{1, ..., r\}$$

with $\tilde{\rho}_k(t) := t^q + \sum_{p=1}^{\infty} c_{p,k} t^{\frac{n+1}{d}+p-1}$.

(II) For all $k \in \{1, ..., r\}$, we define

$$\varphi\left(t^q\right) + \psi\left(\tilde{\rho}_k(t) - t^q\right) = \sum_{l=1}^{qn} d_{l,k} t^{-l} + \sum_{l=0}^{\infty} \tilde{d}_{l,k} t^l =: \tilde{\alpha}_k(t) + \tilde{\delta}_k(t).$$

(III) Let $\lambda_k(t) \in t\mathbb{C}[\![t]\!]$ with $\lambda_k'(0) \neq 0$ and $(\tilde{\rho}_k \circ \lambda_k)(t) = t^q$. We denote by $\alpha_k(t)$ resp. $\delta_k(t)$ the meromorphic resp. holomorphic part of $(\tilde{\alpha}_k + \tilde{\delta}_k) \circ \lambda_k$ and cancel redundant pairs $(\rho, \alpha_k + \delta_k)$ in the sense of Theorem 3.3.2.5. Then we omit the holomorphic parts and receive (after a suitable renumbering of the α_k) the minimal set of parametrizations

$$\left\{(\rho, \alpha_k) | k = 1, ..., r^*\right\}.$$

With $\Gamma = \biguplus_{k=1}^{h} \Gamma_k = \biguplus_{k=1}^{h} \{\alpha \in \Gamma | \exists \xi \in \mathbb{C} : \xi^q = 1 \wedge \alpha(\xi t) = \alpha_k(t)\}$ (see p. 55), we get (after another suitable renumbering of the α_k) the decomposition

$$\left(\mathrm{El}(\mathrm{Id}, \varphi, R) *_{+0} \mathrm{El}\left(\mathrm{Id}, \psi, S\right)\right)_0^{\wedge} \cong \bigoplus_{k=1}^{h} \mathrm{El}\left(\left[t \mapsto t^q\right], \alpha_k, T_{\alpha_k}\right)$$

with regular connections T_{α_k} which all have the same rank. □

4.1.5 Remark: In the setting of Theorem 4.1.4, we have the equivalent decomposition

$$\left(\mathrm{El}(\mathrm{Id}, \varphi, R) *_{+0} \mathrm{El}\left(\mathrm{Id}, \psi, S\right)\right)_0^{\wedge} \cong \bigoplus_{k=1}^{h} \mathrm{El}\left(\tilde{\rho}_k, \tilde{\alpha}_k, T_{\alpha_k}\right)$$

due to [1, Lemma 2.2].

4.1.6 Example: Let

$$\varphi(x) = \frac{1}{2x^2} \text{ and } \psi(y) = \frac{1}{y}.$$

We have deduced that the set C' from page 53 has to be parameterized.

$$\varphi'(x) - \psi'(y) = -\frac{1}{x^3} + \frac{1}{y^2} = 0.$$

Applying the direct image γ_+ with $\gamma : (x, y) \mapsto (x, x + y) =: (u, z)$ gives

$$-\frac{1}{u^3} + \frac{1}{(z-u)^2} = 0 \Leftrightarrow u^3 - (z-u)^2 = 0.$$

As $\gcd(3, 2) = 1$, the polynomial just has one irreducible component (see Remark 3.3.2.2(2)). Thus it is enough to continue with the parametrization $z(u) := u + u^{\frac{3}{2}}$. Using $\beta : t \mapsto t^2 = u$, we get

$$\tilde{\rho}(t) := t^2 + t^3.$$

Now let $\lambda(t) \in t\mathbb{C}[\![t]\!]$ with $\lambda'(0) \neq 0$ and $(\tilde{\rho} \circ \lambda)(t) = t^2$. Moreover we have

$$\tilde{\alpha}(t) := \varphi\left(t^2\right) + \psi\left(\tilde{\rho}(t) - t^2\right) = \frac{1}{2t^4} + \frac{1}{t^3}.$$

Define $\alpha := \tilde{\alpha} \circ \lambda \mod \mathbb{C}[\![t]\!]$. So this gives the decomposition

$$\left(\mathrm{El}\left(\mathrm{Id}, \frac{1}{2x^2}, R\right) *_{+0} \mathrm{El}\left(\mathrm{Id}, \frac{1}{y}, S\right)\right)^{\wedge}_{0} \cong \mathrm{El}\left(\left[t \mapsto t^2\right], \alpha, T_\alpha\right).$$

4.1.7 Example: Let

$$\varphi(x) = \frac{1}{3x^3} \text{ and } \psi(y) = \frac{1}{2y^2}.$$

The same computations as in the previous example give

$$\tilde{\rho}_k(t) = t^3 + \zeta_3^k t^4, \ \tilde{\alpha}_k(t) = \frac{1}{3t^9} + \frac{\zeta_3^k}{2t^8}, \ k \in \{1, 2, 3\}.$$

But as the polynomial which defines the set C' just has $\gcd(4, 3) = 1$ irreducible component, it is enough to continue with one $k \in \{1, 2, 3\}$. With Remark 4.1.5, this gives

$$\left(\mathrm{El}\left(\mathrm{Id}, \frac{1}{3x^3}, R\right) *_{+0} \mathrm{El}\left(\mathrm{Id}, \frac{1}{2y^2}, S\right)\right)^{\wedge}_{0} \cong \mathrm{El}\left(\tilde{\rho}_k, \tilde{\alpha}_k, T_{\tilde{\alpha}_k}\right) \ \forall k \in \{1, 2, 3\}.$$

4.2. The case $n = m$

We have see on page 53 that in this case, the parametrizations of the set

$$C' = \left\{ (u,z) \middle| \sum_{i=1}^{n} i a_i u^{-i-1} - \sum_{j=1}^{m} j b_j (z-u)^{-j-1} = 0 \right\}$$

have the form

$$(u, z(u)) = \left(u, u + \sum_{p=1}^{\infty} c_p u^p \right)$$

and that the order of the series $z(u)$ increases if $c_1 = -1$. We know from the proof of Lemma 4.2 that $c_1 = \zeta_{m+1} \sqrt[m+1]{\frac{mb_m}{na_n}}$. So $c_1 = -1$ is possible if and only if the equation $(-1)^{n+1} a_n = b_n$ holds. But we will show (Corollary 4.2.7) that more terms may vanish - it is even possible that $z(u) = 0$ is a parametrization of C'. That is why we define

$$\varrho := \max_i \left\{ (-1)^{n+1} i a_i - (-1)^{n-i} i b_i \neq 0 \right\} = \max_i \left\{ (-1)^i a_i + b_i \neq 0 \right\}.$$

So in terms of ϱ, the series $z(u)$ is of order $n - \varrho + 1$ if $c_1 = -1$.
If $\varrho > -\infty$, i.e., the set $\{(-1)^i a_i + b_i \neq 0\}$ is not empty, there are $n+1$ parametrizations and none of them equals zero.
If $\varrho = -\infty$, $z = 0$ also is a parametrization of the C'. But such a parametrization is of no interest for us. That is why we can assume $\varrho > -\infty$ without loss of generality.
Now we want to compute the parametrizations $z_k(u)$ for all k. In order to find all parametrizations $z_k(u)$, we consider the two cases

$$\varrho = n \text{ and } \varrho \in \{0, ..., n-1\}.$$

The case $\varrho = n$

Corollary 4.2.1: The set C' (see page 53) has exactly $n+1$ parametrizations of the form

$$(u, z_k(u)) = \left(u, u + \sum_{p=1}^{\infty} c_{p,k} u^p \right), \quad k \in \{1, ..., n+1\}.$$

Proof: This is also an immediate consequence of Lemma 4.2. As $\varrho = n$, we have $(-1)^{n+1} a_n \neq b_n$. Hence $c_{1,k} = \zeta_{n+1}^k \sqrt[n+1]{\frac{b_n}{a_n}} \neq -1$ for all $k \in \{1, ..., n+1\}$. This means, all parametrizations $z_k(u)$ are of order 1. □

As the parametrizations $z_k(u)$ already have integer exponents, we can set $\tilde{\rho}_k(u) := z_k(u)$ for all k and it is $P(u,z) = \prod_{k=1}^{n+1}(z - \tilde{\rho}_k(u))$ a decomposition in pairwise distinct irreducible components. Now let $\lambda_k(u) \in u\mathbb{C}[\![u]\!]$ with $\lambda_k'(0) \neq 0$ and $\rho(u) := (\tilde{\rho}_k \circ \lambda_k)(u) = u$. Moreover we have

$$\varphi(\lambda_k(u)) + \psi(\rho(u) - \lambda_k(u)) = \sum_{i=1}^{n} a_i \frac{1}{\lambda_k(u)^i} + \sum_{i=1}^{n} b_i \frac{1}{(u - \lambda_k(u))^i}, \ k \in \{1, ..., n+1\}.$$

It is easy to see that $\varphi(\lambda_k(u))$ and $\psi(u - \lambda_k(u))$ both have the same pole order n. This means, after polynomial division, we receive the Laurent series

$$\underbrace{\sum_{l=1}^{n} d_{l,k} u^{-l}}_{=:\alpha_k(u) \text{ (meromorphic part)}} + \underbrace{\sum_{l=0}^{\infty} \tilde{d}_{l,k} u^{l}}_{=:\delta_k(u) \text{ (holomorphic part)}} , k \in \{1, ..., n+1\}.$$

Next we have to cancel redundant pairs in the sense of Theorem 3.3.2.5 again. This means, as $\rho = \mathrm{Id}$, a minimal representation of the singular locus $\mathscr{S}$ consists of r^* pairs $(\mathrm{Id}, \alpha_k + \delta_k)$ with $r^* \leq n+1$ and $r^* < n+1$ if and only if **equal** pairs occur.
Due to $\rho = \mathrm{Id}$, the application of Theorem 3.3.2.4 now gives the decomposition

$$(\mathrm{El}\,(\mathrm{Id}, \varphi, R) *_{+0} \mathrm{El}\,(\mathrm{Id}, \psi, S))_0^{\wedge} \cong \bigoplus_{k=1}^{h} \mathscr{E}^{\alpha_k} \otimes T_{\alpha_k}.$$

with $h \leq r^*$ and $h < r^*$ if and only if not all $\alpha_k, k \in \{1, ..., r^*\}$, are pairwise distinct.
Now let us compute

$$\Psi_\eta \left(\mathscr{E}^{\varphi(u) + \psi(\rho(\eta) - u) - \alpha_k(\eta)} \otimes (R \boxtimes S) \right).$$

We have

$$\varphi(u) + \psi(\rho(\eta) - u) - \alpha_k(\eta) = \frac{1}{u^n \eta^n} \left(\eta^n a(u) + u^n \eta^n \sum_{i=1}^{n} b_i \frac{1}{(\eta - u)^i} - u^n d_k(\eta) \right),$$

where $a(u) := \sum_{i=1}^{n} a_i u^{n-i}$ and $d_k(\eta) := \sum_{l=1}^{n} d_{l,k} \eta^{n-l}$ are holomorphic functions.
Now, in order to solve the singularity at $(0,0)$, we use blow up the ideal (u, η) again. It is enough to consider the chart $e_0 : (v, \eta) \mapsto (v\eta, \eta)$ (this can be proven with the same method as in Lemma 3.3.2.6 resp. Lemma 4.1.2). So the application of e_0^+ gives

$$\frac{1}{v^n \eta^{2n}} \left(\eta^n a(v\eta) + v^n \eta^{2n} \sum_{i=1}^{n} b_i \frac{1}{(\eta - v\eta)^i} - v^n \eta^n d_k(\eta) \right)$$

$$=\frac{1}{v^n\eta^n}\left(a(v\eta)+v^n\sum_{i=1}^{n}b_i\eta^{n-i}\frac{1}{(1-v)^i}-v^n d_k(\eta)\right)$$
$$=\frac{1}{v^n\eta^n(1-v)^n}\left((1-v)^n a(v\eta)+v^n\sum_{i=1}^{n}b_i\eta^{n-i}-(1-v)^n v^n d_k(\eta)\right)=:\frac{f_k(v,\eta)}{v^n\eta^n(1-v)^n}$$

Note that in analogy to the earlier cases, the other chart $e_1:(v,\eta)\mapsto(v,v\eta)$ of the blowing-up space has no influence on our computations. So we can equivalently determine

$$\Psi_\eta\left(\mathscr{E}^{f_k(v,\eta)/v^n(1-v)^n\eta^n}\otimes e_0^+(R\boxtimes S)\right).$$

instead of

$$\Psi_\eta\left(\mathscr{E}^{\varphi(u)+\psi(\rho(\eta)-u)-\alpha_k(\eta)}\otimes(R\boxtimes S)\right).$$

Now let us consider the support $\{f_k(v,0)=0\}$ again:

$$f_k(v,0)=0\Leftrightarrow(1-v)^n a_n+v^n b_n-v^n(1-v)^n d_{n,k}=0.$$

Unfortunately we cannot make a general statement about the existence of double roots of $f_k(v,0)$. But we will show in the following lemma that $f_k(v,0)$ has no roots of higher multiplicity than 2. The result holds even if a_n, b_n and $d_{n,k}$ are independent from each other (which is not the case here, as $d_{n,k}$ is an expression in a_n and b_n). Note that $f_k(1,\eta)\neq 0$. That is why the denominator of the expression $\frac{f_k(v,\eta)}{v^n\eta^n(1-v)^n}$ makes no problems here.

4.2.2 Lemma: The polynomial

$$g(v):=a(1-v)^n+bv^n-dv^n(1-v)^n\in\mathbb{C}[v]$$

with $a,b,d\in\mathbb{C}\setminus\{0\}$ has only roots of multiplicity ≤ 2.

Proof: We assume that $\bar{v}$ is a root of $g(v)$ with $(v-\bar{v})^3|g(v)$. Then we have

$$\begin{aligned}
g(\bar{v})&=a(1-\bar{v})^n+b\bar{v}^n-d\bar{v}^n(1-\bar{v})^n=0\\
g'(\bar{v})&=-na(1-\bar{v})^{n-1}+nb\bar{v}^{n-1}+nd\bar{v}^n(1-\bar{v})^{n-1}-nd\bar{v}^{n-1}(1-\bar{v})^n=0\\
g''(\bar{v})&=(n-1)na(1-\bar{v})^{n-2}+(n-1)nb\bar{v}^{n-2}-(n-1)nd\bar{v}^n(1-\bar{v})^{n-2}\\
&\quad+2n^2d\bar{v}^{n-1}(1-\bar{v})^{n-1}-(n-1)nd\bar{v}^{n-2}(1-\bar{v})^n=0.
\end{aligned}$$

Note that $\bar{v}\in\{0,1\}$ is not possible as $g(0)\neq 0$ and $g(1)\neq 0$. The first equation $g(\bar{v})=0$ is equivalent to

$$d\bar{v}^n(1-\bar{v})^n=a(1-\bar{v})^n+b\bar{v}^n. \tag{4.2}$$

Together with the equation $g'(\bar{v}) = 0$, this gives

$$b\bar{v}^n = \frac{1}{\bar{v}} a(1-\bar{v})^{n+1}. \tag{4.3}$$

Then (4.2) and (4.3) in the equation $g''(\bar{v}) = 0$ imply

$$\begin{aligned}
&(n-1)a(1-\bar{v})^{n-2} + (n-1)b\bar{v}^{n-2} + d\bar{v}^n(1-\bar{v})^n \left(\frac{1-n}{\bar{v}^2} + \frac{2n}{\bar{v}(1-\bar{v})} - \frac{n-1}{(1-\bar{v})^2}\right) = 0\\
\Leftrightarrow &(n-1)\left(a\bar{v}^2(1-\bar{v})^n + b\bar{v}^n(1-\bar{v})^2\right)\\
&\quad + d\bar{v}^n(1-\bar{v})^n\left((1-n)(1-\bar{v})^2 + 2n\bar{v}(1-\bar{v}) + (1-n)\bar{v}^2\right) = 0\\
\Leftrightarrow &\left(n\bar{v}^2 - \bar{v}^2\right) a(1-\bar{v})^n + (n-1)(1-\bar{v})^2 \frac{1}{\bar{v}} a(1-\bar{v})^{n+1}\\
&\quad + \left(a(1-\bar{v})^n + b\bar{v}^n\right)\left(2\bar{v}^2 - 4n\bar{v}^2 + 4n\bar{v} - 2\bar{v} - n + 1\right) = 0\\
\Leftrightarrow &n\bar{v}^2 - \bar{v}^2 + (n-1)(1-\bar{v})^3\frac{1}{\bar{v}} + \left(1 + \frac{1-\bar{v}}{\bar{v}}\right)\left(2\bar{v}^2 - 4n\bar{v}^2 + 4n\bar{v} - 2\bar{v} - n + 1\right) = 0\\
\Leftrightarrow &n\bar{v}^3 - \bar{v}^3 + (n-1)(1-\bar{v})^3 + 2\bar{v}^2 - 4n\bar{v}^2 + 4n\bar{v} - 2\bar{v} - n + 1 = 0\\
\Leftrightarrow &-3n\bar{v} + 3n\bar{v}^2 + 3\bar{v} - 3\bar{v}^2 + 2\bar{v}^2 - 4n\bar{v}^2 + 4n\bar{v} - 2\bar{v} = 0\\
\Leftrightarrow &-\bar{v} - n\bar{v} + 1 + n = 0 \Leftrightarrow \bar{v} = 1.
\end{aligned}$$

But we have already shown that $\bar{v} = 1$ is not possible. Thus we have a contradiction. □

In order to get some information about the **existence** of double roots, we will again compute the regular part with respect to $\tilde{\alpha}_k$, but this time we use the original setting

$$\varphi(\eta) + \psi(\tilde{\rho}_k(\eta) - \eta)$$

instead of the representation $\sum_{l=1}^{n} d_{l,k} u^{-l}$ or $\tilde{\alpha}_k \circ \lambda_k$. Again, this works even in consideration of the holomorphic terms of $\varphi(\eta) + \psi(\tilde{\rho}_k(\eta) - \eta)$, hence we include them. Then the same computations as above (including the same blowup) give $\frac{\tilde{f}_k(v,\eta)}{v^n \eta^n}$ with

$$\tilde{f}_k(v,\eta) = \sum_{i=1}^{n} a_i \eta^{n-i}\left(v^{n-i} - v^n\right) + \sum_{i=1}^{n} b_i v^n \eta^{n-i}\left(\left(\frac{\tilde{\rho}_k(\eta)}{\eta} - v\right)^{-i} - \left(\frac{\tilde{\rho}_k(\eta)}{\eta} - 1\right)^{-i}\right).$$

Having $\tilde{f}_k(v,\eta)$ in this form, we see that $v-1$ divides $\tilde{f}_k(v,\eta)$. This is no contradiction to the results of the other computation of the regular part from above (where we have shown that $f_k(1,\eta) \neq 0$) because the roots change depending on the usage of the function $\lambda_k(u)$ with $\lambda_k'(0) \neq 0$ and $\tilde{\rho}(\lambda_k(u)) = u$.
Further computations show that $(v-1)^2$ divides $\tilde{f}_k(v,\eta)$ because the expression $\frac{\partial}{\partial v}\tilde{f}_k(1,\eta)$

fulfills the equation defining the set C' (page 53) just as in the case $m < n$. That $(v-1)^l$ with $l > 2$ does not divide $f_k(v,\eta)$, is a consequence of Lemma 4.2.2 (of course, the multiplicities of the roots of $f_k(v,\eta)$ do not depend on the usage of $\lambda_k(u)$).
Thus we have shown that there exists at least one $\bar{v}$ with $(v-\bar{v})^2 | \tilde{f}_k(v,\eta)$. So again, the final result is that **every** regular part in the decomposition $\mathscr{E}^{\alpha_k} \otimes T_{\alpha_k}$ of the local formal convolution $(\mathrm{El}(\mathrm{Id},\varphi,R) *_{+_0} \mathrm{El}(\mathrm{Id},\psi,S))_0^\wedge$ is not zero. The existence of other double roots cannot be proved with our methods. It varies from case to case. However, this is enough for us to give the next theorem:

4.2.3 Theorem: Let $\varphi(x) = \sum_{i=1}^n a_i x^{-i}$ resp. $\psi(y) = \sum_{i=1}^m b_i y^{-i}$ be meromorphic functions with pole orders $\boldsymbol{n} = \boldsymbol{m}$ and $(-1)^n a_n + b_n \neq 0$. Then

(I) After applying the direct image γ_+ with $\gamma : (x,y) \mapsto (x, x+y) =: (u,z)$ to the set $C = \{(x,y) | \varphi'(x) = \psi'(y)\}$, the resulting set C' (see page 53) has Puiseux parametrizations of the form

$$(u, \tilde{\rho}_k(u)) = \left(u, u + \sum_{p=1}^{\infty} c_{p,k} u^p\right), \quad k \in \{1, ..., n+1\}$$

with $u + c_{1,k}u \neq 0$ for all $k = 1, ..., n+1$.

(II) For all $k \in \{1, ..., n+1\}$, we define

$$\varphi(u) + \psi(\tilde{\rho}_k(u) - u) = \sum_{l=1}^{n} d_{l,k} u^{-l} + \sum_{l=0}^{\infty} \tilde{d}_{l,k} u^l =: \tilde{\alpha}_k(u) + \tilde{\delta}_k(t).$$

(III) Let $\lambda_k(u) \in u\mathbb{C}[\![u]\!]$ with $\lambda_k'(0) \neq 0$ and $\rho := (\tilde{\rho}_k \circ \lambda_k)(u) = u$. We denote by $\alpha_k(t)$ resp. $\delta_k(t)$ the meromorphic resp. holomorphic part of $\left(\tilde{\alpha}_k + \tilde{\delta}_k\right) \circ \lambda_k$ and cancel redundant pairs $(\rho, \alpha_k + \delta_k)$, i.e., pairs which occur more than once. Then we receive a minimal representation of the singular locus $\mathscr{S}$ consisting of $r^* \leq n+1$ pairwise distinct pairs $(\rho, \alpha_k + \delta_k)$. Next we omit the holomorphic parts δ_k and cancel pairs $(\rho, \alpha_k), k \in \{1, ..., r^*\}$, which occur more than once. Let $(\rho, \alpha_k), k = 1, ..., h$, be the pairs left (after a suitable renumbering).

Then we get the decomposition

$$(\mathrm{El}(\mathrm{Id},\varphi,R) *_{+_0} \mathrm{El}(\mathrm{Id},\psi,S))_0^\wedge \cong \bigoplus_{k=1}^{h} \mathrm{El}(\mathrm{Id},\alpha_k,T_{\alpha_k})$$

with regular connections T_{α_k}. □

4.2.4 Example: Let

$$\varphi(x) = -\frac{1}{x} \text{ and } \psi(y) = \frac{1}{y}.$$

The set C' (see page 53) has to be parameterized now. We have

$$\varphi'(x) - \psi'(y) = \frac{1}{x^2} + \frac{1}{y^2} = 0.$$

Applying the direct image under $(x, y) \mapsto (x, x + y) =: (u, z)$, we get the equation

$$\frac{1}{u^2} + \frac{1}{(z-u)^2} = 0 \Leftrightarrow u^2 + (z-u)^2 = 0$$

which defines C'. So it is easy to see that $\tilde{\rho}_\pm(u) := z_\pm(u) = (1 \pm i)u$ are the corresponding parametrizations. The next step is the computation of $\alpha_\pm(u)$. First of all let

$$\tilde{\alpha}_\pm(u) := \varphi(u) + \psi(\tilde{\rho}_\pm(u) - u) \in u^{-1}\mathbb{C}[u^{-1}].$$

Using $\lambda_\pm(u) := \frac{1}{1 \pm i} u$, we get

$$\begin{aligned} \rho(u) :=& \tilde{\rho}_\pm(\lambda_\pm(u)) = u \\ \alpha_\pm(u) :=& \tilde{\alpha}_\pm(\lambda_\pm(u)) = \varphi(\lambda_\pm(u)) + \psi(u - \lambda_\pm(u)) = \mp\frac{2i}{u}. \end{aligned}$$

Note that, as $\alpha_+ \neq \alpha_-$, we cannot cancel one of the pairs $(\rho, \alpha_+), (\rho, \alpha_-)$.
In order to compute the regular part with respect to $\alpha_\pm(\eta)$, we now have to consider the module

$$\Psi_\eta\left(\mathscr{E}^{\varphi(u)+\psi(\eta-u)-\alpha_\pm(\eta)} \otimes (R \boxtimes S)\right).$$

We have

$$\varphi(u) + \psi(\eta - u) - \alpha_\pm(\eta) = -\frac{1}{u} + \frac{1}{\eta - u} \pm 2i\frac{1}{\eta} = \frac{1}{u\eta}\left(-\eta + \frac{u\eta}{\eta - u} \pm 2iu\right).$$

In order to solve the singularity, we need one blowup of the form $(v, \eta) \mapsto (v\eta, \eta)$. So we consider the chart $e_0 : (v, \eta) \mapsto (v\eta, \eta)$ and the polynomial $f_\pm(v, \eta)$, where

$$\begin{aligned} &\frac{1}{v\eta^2}\left(-\eta + \frac{v\eta^2}{\eta - v\eta} \pm 2iv\eta\right) = \frac{1}{v\eta}\left(-1 + \frac{v}{1-v} \pm 2iv\right) \\ =&\frac{1}{v(1-v)\eta}\left(-(1-v) + v \pm 2iv(1-v)\right) =: \frac{f_\pm(v,\eta)}{v(1-v)\eta}. \end{aligned}$$

It is easy to see that $f(v, \eta) = f(v, 0)$ has the double root $v = \frac{1 \mp i}{2}$. So we get the decomposition

$$(\mathrm{El}(\mathrm{Id}, \varphi, R) *_{+_0} \mathrm{El}(\mathrm{Id}, \psi, S))_0^\wedge \cong \mathrm{El}(\mathrm{Id}, \alpha_+, T_+) \oplus \mathrm{El}(\mathrm{Id}, \sigma_-, T_-).$$

4.2.5 Example: Let

$$\varphi(x) = \frac{1}{2x^2} \text{ and } \psi(y) = \frac{1}{2y^2}.$$

Applying γ_+ with $\gamma : (x, y) \mapsto (x, x+y) =: (u, z)$ to the set

$$C = \{(x, y) | \varphi'(x) - \psi'(y) = 0\},$$

we receive the set C' which we have to parameterize now. We have

$$\varphi'(x) - \psi'(y) = -\frac{1}{x^3} + \frac{1}{y^3} = 0.$$

With γ_+, we receive the equation

$$-\frac{1}{u^3} + \frac{1}{(z-u)^3} = 0 \Leftrightarrow u^3 - (z-u)^3 = 0.$$

Thus $z_k(u) = \tilde{\rho}_k(u) = \left(1 + \zeta_3^k\right) u, k \in \{1, 2, 3\}$, are the parametrizations. The next step is the computation of the $\alpha_k(u)$. We have

$$\tilde{\alpha}_k(u) := \varphi(u) + \psi(\tilde{\rho}_k(u) - u) = \left(1 + \zeta_3^k\right) \frac{1}{2u^2}, k \in \{1, 2, 3\}.$$

Using $\lambda_k(u) := \frac{1}{1+\zeta_3^k} u$, we get

$$\rho(u) := \tilde{\rho}_k(\lambda_k(u)) = u \text{ and } \alpha_k(u) := \tilde{\alpha}_k(\lambda_k(u)) = (1 + \zeta_3^k)^3 \frac{1}{2u^2} \ \forall k.$$

As $(1 + \zeta_3^k)^3 = -1$ for $k = 1$ **and** $k = 2$, we cancel the redundant pair (ρ, α_1). Hence the singular locus is parameterized by the $h = 2$ curves (ρ, α_2) and (ρ, α_3). In order to check the regular part with respect to $\alpha_k(\eta)$, $k \in \{2, 3\}$, we consider

$$\Psi_\eta \left(\mathscr{E}^{\varphi(u) + \psi(\eta - u) - \alpha_k(\eta)} \otimes (R \boxtimes S) \right).$$

Now

$$\varphi(u) + \psi(\eta - u) - \alpha_k(\eta)) = \frac{1}{u^2 \eta^2} \left(\frac{1}{2} \eta^2 + \frac{u^2 \eta^2}{2(\eta - u)^2} - (1 + \zeta_3^k)^3 \frac{1}{2} u^2 \right).$$

In order to solve the singularity, we need one blowup of the form $(v,\eta) \mapsto (v\eta,\eta)$. We get

$$\begin{aligned}
&\frac{1}{v^2\eta^4}\left(\frac{1}{2}\eta^2 + \frac{v^2\eta^4}{2(\eta - v\eta)^2} - \frac{1}{2}(1+\zeta_3^k)^3 v^2\eta^2\right)\\
=&\frac{1}{v^2\eta^2}\left(\frac{1}{2} + \frac{v^2}{2(1-v)^2} - \frac{1}{2}(1+\zeta_3^k)^3 v^2\right)\\
=&\frac{1}{2v^2(1-v)^2\eta^2}\left((1-v)^2 + v^2 - (1+\zeta_3^k)^3 v^2(1-v)^2\right) =: \frac{f_k(v,\eta)}{2v^2(1-v)^2\eta^2}
\end{aligned}$$

with polynomials

$$\begin{aligned}
f_2(v,\eta) &= f_2(v) = 0 \Leftrightarrow (v^2 - v + 1)^2 = 0,\\
f_3(v,\eta) &= f_3(v) = 0 \Leftrightarrow -8v^4 + 16v^3 - 6v^2 - 2v + 1 = 0.
\end{aligned}$$

Hence $f_2(v,\eta)$ has exactly two double roots which are the two primitive sixth roots of unity. f_3 has exactly one double root at $v = \frac{1}{2}$. This gives the following decomposition:

$$\begin{aligned}
(\mathrm{El}\,(\mathrm{Id},\varphi,R) *_{+0} \mathrm{El}\,(\mathrm{Id},\psi,S))_0^{\wedge} &\cong \bigoplus_{k=2}^{3} \mathrm{El}\,(\tilde{\rho}_k(u),\tilde{\alpha}_k(u),T_{\tilde{\alpha}_k})\\
&\cong \bigoplus_{k=2}^{3} \mathrm{El}\,((\tilde{\rho}_k \circ \lambda_k)(u),(\tilde{\alpha}_k \circ \lambda_k)(u),T_\alpha)\\
&= \bigoplus_{k=2}^{3} \mathrm{El}\,(\rho(u),\alpha(u),T_\alpha)\\
&= \bigoplus_{k=2}^{3} \mathrm{El}\,([u \mapsto u],\alpha(u),T_\alpha)\\
&= \mathrm{El}\left(\mathrm{Id}, -\frac{1}{2u^2}, T_{\alpha_2}\right) \oplus \mathrm{El}\left(\mathrm{Id}, \frac{4}{u^2}, T_{\alpha_3}\right).
\end{aligned}$$

This means, the representation $\bigoplus_{k=1}^{3} \mathrm{El}(\tilde{\rho}_k,\tilde{\alpha}_k,T_{\tilde{\alpha}_k})$ does not have the minimal number of direct summands because the meromorphic functions α_k are equal for $k = 1$ and $k = 2$. So the convolution is isomorphic to a direct sum of two elementary connections. But the rank of the regular part has to be adapted to the new situation, of course.

The case $\varrho < n$

4.2.7 Corollary: The set C' (see page 53) has exactly n parametrizations

$$(u, \tilde{\rho}_k(u)) := \left(u, u + \sum_{p=1}^{\infty} c_{p,k} u^p\right), \quad k = 1, ..., n,$$

of order 1, and exactly one parametrization of order $n - \varrho + 1$ which is given by

$$(u, \tilde{\rho}_{n+1}(u)) := \left(u, \sum_{p=n-\varrho+1}^{\infty} c_{p,n+1} u^p\right).$$

Proof: Here we have to repeat the first parametrization steps of Lemma 4.2. We want to parameterize the polynomial equation $P(x, y) = 0$ with

$$P(x,y) = nb_n x^{n+1} + (n-1)b_{n-1}x^{n+1}y + ... + b_1 x^{n+1} y^{n-1} + \left(\sum_{i=1}^{n} -ia_i x^{n-i}\right) y^{n+1}.$$

The Newton polygon of this equation has only one slope which has 1 as absolute value. Hence we have to set $y = c_1 x + \tilde{y}_1$ and consider the function $P(x, c_1 x + \tilde{y}_1)$. We will only need the coefficient of $\tilde{y}_1^0$ because all other results follow from Lemma 4.2. The coefficient of $\tilde{y}_1^0$ is the polynomial

$$nb_n x^{n+1} + (n-1)b_{n-1}c_1 x^{n+2} + ... + b_1 c_1^{n-1} x^{2n} + c_1^{n+1} \sum_{i=1}^{n} -ia_i x^{2n-i+1}$$

$$= \sum_{i=1}^{n} \left(ib_i c_1^{n-i} - ia_i c_1^{n+1}\right) x^{2n-i+1}$$

Choosing c_1 such that the term $\left(nb_n - na_n c_1^{n+1}\right) x^{n+1}$ vanishes, leads to the well-known values $c_{1,k} = \zeta_{n+1}^k \sqrt[n+1]{\frac{b_n}{a_n}}$, $k \in \{1, ..., n+1\}$. By definition, $\varrho < n$, hence $(-1)^n a_n + b_n = 0$. So

$$c_{1,k} = \zeta_{n+1}^k \sqrt[n+1]{\frac{b_n}{a_n}} = \zeta_{n+1}^k \sqrt[n+1]{\frac{(-1)^{n+1} a_n}{a_n}} = -\zeta_{n+1}^k, \ k \in \{1, ..., n+1\}.$$

Then the set C' has parametrizations of the form

$$(u, z_k(u)) = \left(u, u - \zeta_{n+1}^k u + \sum_{p=2}^{\infty} c_p u^p\right), \quad k \in \{1, ..., n+1\}.$$

Now it is easy to see that, for $k = 1, ..., n$, the parametrizations z_k have order 1, hence the desired form of the corollary.
So let us consider the case $k = n + 1$, i.e., $c_1 = -1$. Then the coefficient of $\tilde{y}_1^0$ in the polynomial $P(x, c_1 x + \tilde{y}_1)$ has the form

$$\sum_{i=1}^{n} \left(ib_i(-1)^{n-i} - ia_i(-1)^{n+1}\right) x^{2n-i+1}$$

with

$$ib_i(-1)^{n-i} - ia_i(-1)^{n+1} = 0 \Leftrightarrow ib_i - ia_i(-1)^{n+1-(n-i)} = 0 \Leftrightarrow ib_i - ia_i(-1)^{i+1} = 0$$
$$\Leftrightarrow ib_i + ia_i(-1)^i = 0.$$

By definition, $\varrho = \max_i\{ib_i + ia_i(-1)^i \neq 0\}$. This means, the term

$$\left(\varrho b_\varrho(-1)^{n-\varrho} - \varrho a_\varrho(-1)^{n+1}\right) x^{2n-\varrho+1} \neq 0$$

is the term containing the lowest x-exponent in the coefficient of $\tilde{y}_1^0$. Then the proof of Lemma 4.2 shows that the next Newton polygon has two slopes with absolute values $n - \varrho + 1$ and 1. □

Note that, as in the case $\varrho = n$, the defining polynomial of the set C' (see page 53) decomposes in a product of $r = n + 1$ pairwise distinct irreducible components. So firstly, we do not have to cancel any parametrization.
To complete the parametrization of the singular locus, we have to compute the meromorphic functions

$$\varphi(u) + \psi(\tilde{\rho}_k(u) - u) = \sum_{i=1}^{n} a_i \frac{1}{u^i} + \sum_{i=1}^{n} b_i \frac{1}{(\tilde{\rho}_k(u) - u)^i}, \quad k \in \{1, ..., n+1\}$$

which have pole order n for $k \in \{1, ..., n\}$. An easy computation shows that for $k = n+1$, we have pole order ϱ because $(-1)^i a_i + b_i = 0$ for $i = \varrho + 1, ..., n$.
Next we use $\lambda_k(u) \in u\mathbb{C}[\![u]\!]$ with $\lambda'(0) \neq 0$ and

$$\rho_k(u) := (\tilde{\rho}_k \circ \lambda_k)(u) = \begin{cases} u & \text{if } k \in \{1, ..., n\}, \\ u^{n-\varrho+1} & \text{if } k = n + 1. \end{cases}$$

Moreover we get the functions

$$\varphi(\lambda_k(u))+\psi(\rho_k(u)-\lambda_k(u))=\begin{cases}\sum_{l=1}^{n} d_{l,k}u^{-l}+\sum_{l=0}^{\infty}\tilde{d}_{l,k}u^{l} & \text{if } k\in\{1,...,n\},\\ \sum_{l=1}^{\varrho} d_{l,n+1}u^{-l}+\sum_{l=0}^{\infty}\tilde{d}_{l,n+1}u^{l} & \text{if } k=n+1.\end{cases}$$

As in the last sections, we denote by $\alpha_k(u)$ the meromorphic part of the series, and by $\delta_k(u)$ the holomorphic part. Note that the pole order of α_k is equal to the pole order of $\tilde{\alpha}_k$ for all k. Now we are able to determine a minimal representation $\mathscr{S}$ of the singular locus. We consider the functions $\alpha_k(u)+\delta_k(u), k\in\{1,...,n\}$, and cancel the redundant ones, i.e., the functions $\alpha_k(u)+\delta_k(u)$ with $\alpha_k(\zeta u)+\delta_k(\zeta u)=\alpha_l(u)+\delta_l(u)$ for a ζ with $\zeta^{n-\varrho+1}=1$ and $l\in\{1,...,n\}$ with $k\neq l$. Omitting the holomorphic parts δ_k, we get (for a suitable renumbering) a set of pairs

$$\{(\mathrm{Id},\alpha_k(u))\,|k=1,...,r^*,r^*\leq n\}\cup\left\{\left(\left[u\mapsto u^{n-\varrho+1}\right],\alpha_{n+1}(u)\right)\right\}$$

which is minimal in the sense that the equality $\alpha_k(\zeta u)+\delta_k(\zeta u)=\alpha_l(u)+\delta_l(u)$ never holds for arbitrary $k,l\in\{1,...,r^*\}$ with $k\neq l$ and ζ with $\zeta^{n-\varrho+1}=1$.
Then after Theorem 3.3.2.4, with $\rho(u):=u^{n-\varrho+1}$, the convolution looks like this:

$$\left(\rho^+\left(\mathrm{El}(\mathrm{Id},\varphi(x),R)*_{+_0}\mathrm{El}(\mathrm{Id},\psi(y),S)\right)\right)_0^{\wedge}\cong\bigoplus_{\alpha\in\Gamma}\mathscr{E}^{\alpha}\otimes T_{\alpha}$$

with $\Gamma=\left\{\alpha_k\left(\xi u^{n-\varrho+1}\right)\right|k=1,...,r^*;\xi^{n-\varrho+1}=1\}\cup\left\{\alpha_{n+1}(\xi u)\,|\xi^{n-\varrho+1}=1\right\}$. The elements $\alpha_k(\xi t)$ with $k\in\{1,...,r^*\}$ and $\xi^{n-\varrho+1}=1$ do not have to be pairwise non-identical, so we decompose Γ (for a suitable renumbering) in $h+1$ pairwise disjoint subsets with $h\leq r^*$ and

$$\Gamma_k:=\left\{\alpha\in\Gamma\,|\exists\xi\in\mathbb{C}:\xi^{n-\varrho+1}=1\wedge\alpha(\xi t)=\alpha_k(t)\right\}.$$

4.2.8 Proposition: 1) For any $k\in\{1,...,h\}$, the regular parts with respect to the functions

$$\left\{\alpha_k\left(\xi u^{n-\varrho+1}\right)\right|\xi^{n-\varrho+1}=1\}$$

are isomorphic and not equal to zero.

2) For $k=n+1$, the regular parts with respect to the functions

$$\left\{\alpha_{n+1}(\xi u)\,|\xi^{n-\varrho+1}=1\right\}$$

are isomorphic and not equal to zero.

Proof:

1) Consider $k \in \{1, ..., h\}$ and the decomposition

$$\left(\rho^+ \left(\mathrm{El}(\mathrm{Id}, \varphi(x), R) *_{+0} \mathrm{El}(\mathrm{Id}, \psi(y), S)\right)\right)_0^\wedge \cong \bigoplus_{\alpha \in \Gamma} \mathscr{E}^\alpha \otimes T_\alpha$$

from above. We can equivalently write

$$\bigoplus_{\alpha \in \Gamma} \mathscr{E}^\alpha \otimes T_\alpha = \bigoplus_{k=1}^{h} \left(\bigoplus_{\xi^{n-\varrho+1}=1} \mathscr{E}^{\alpha_k(\xi u^{n-\varrho+1})} \otimes T_{\alpha_k} \right) \oplus \left(\bigoplus_{\xi^{n-\varrho+1}=1} \mathscr{E}^{\alpha_{n+1}(\xi u)} \otimes T_\xi \right)$$

and it is obvious (see [1, Remark 2.8]) that for a fixed $k \in \{1, ..., h\}$, the regular parts with respect to $\alpha_k(\xi u^{n-\varrho+1})$ are isomorphic for all ξ with $\xi^{n-\varrho+1} = 1$ because the ramification ρ^+ only has to be applied due to the appearance of the pair $(\rho_{n+1}, \alpha_{n+1})$ as $\rho_k = \mathrm{Id}$ for all $k \in \{1, ..., h\}$. That these regular parts are not equal to zero, follows for the same reasons as in the case $\varrho = n$.

2) Due to 1), the proof reduces to the problem of finding an elementary connection $\mathrm{El}(\eta, \sigma, P)$ with

$$\rho^+ \mathrm{El}(\eta, \sigma, P) \cong \bigoplus_{\xi^{n-\varrho+1}=1} \mathscr{E}^{\alpha_{n+1}(\xi u)} \otimes T_{\alpha_{n+1}(\xi u)}.$$

For this, let $\mathcal{M}$ be a meromorphic connection with

$$\rho^+ \mathcal{M} \cong \bigoplus_{\xi^{n-\varrho+1}=1} \mathscr{E}^{\alpha_{n+1}(\xi u)} \otimes T_{\alpha_{n+1}(\xi u)}. \tag{4.4}$$

Then the refined theorem of Levelt-Turrittin-Malgrange (see the Introduction or [1, Corollary 3.3]) shows that there exists a decomposition

$$\mathcal{M} \cong \bigoplus_i \mathrm{El}(\eta_i, \sigma_i, P_i).$$

This is the unique decomposition with the property that each $\eta_i \mathscr{E}^{\sigma_i}$ is irreducible and $\eta_i \mathscr{E}^{\sigma_i} \not\cong \eta_j \mathscr{E}^{\sigma_j}$ if $i \neq j$. Without loss of generality, we can assume $\eta_i(u) = u^{q_i}$ for all i due to [1, Lemma 2.2]. Moreover, due to the isomorphism (4.4),

$$q_i | (n - \varrho + 1)$$

for all i. Now let P_i^{1/q_i} be a regular connection with $\eta_i^+ P_i^{1/q_i} \cong P_i$. Then

$$\begin{aligned}
\rho^+\mathcal{M} &\cong \rho^+\left(\bigoplus_i \mathrm{El}(\eta_i, \sigma_i, P_i)\right) \cong \bigoplus_i \rho^+\eta_{i,+}\left(\mathscr{E}^{\sigma_i} \otimes P_i\right) \\
&\cong \bigoplus_i \rho^+\eta_{i,+}\left(\mathscr{E}^{\sigma_i} \otimes \eta_i^+ P_i^{1/q_i}\right) \cong \bigoplus_i \left(\rho^+\eta_{i,+}\mathscr{E}^{\sigma_i}\right) \otimes \left(\rho^+ P_i^{1/q_i}\right) \\
&\cong \bigoplus_{\xi^{n-\varrho+1}=1} \mathscr{E}^{\alpha_{n+1}(\xi u)} \otimes T_{\alpha_{n+1}(\xi u)}
\end{aligned}$$

because of (4.4). In consideration of the fact that we have $q_i|(n-\varrho+1)$ for all i and [1, Lemma 2.4],

$$\begin{aligned}
\rho^+\eta_{i,+}\mathscr{E}^{\sigma_i(u)} &= \left[u \mapsto u^{n-\varrho+1}\right]^+ \left[u, \mapsto u^{q_i}\right]_+ \mathscr{E}^{\sigma_i(u)} \\
&\cong \left[u \mapsto (u^{q_i})^{(n-\varrho+1)/q_i}\right]^+ \left[u, \mapsto u^{q_i}\right]_+ \mathscr{E}^{\sigma_i(u)} \\
&\cong \left[u \mapsto u^{(n-\varrho+1)/q_i}\right]^+ \left[u, \mapsto u^{q_i}\right]^+ \left[u, \mapsto u^{q_i}\right]_+ \mathscr{E}^{\sigma_i(u)} \\
&\cong \left[u \mapsto u^{(n-\varrho+1)/q_i}\right]^+ \bigoplus_{\zeta^{q_i}=1} \mathscr{E}^{\sigma_i(\zeta u)} \\
&\cong \bigoplus_{\zeta^{q_i}=1} \left[u \mapsto u^{(n-\varrho+1)/q_i}\right]^+ \mathscr{E}^{\sigma_i(\zeta u)} \\
&\cong \bigoplus_{\zeta^{q_i}=1} \mathscr{E}^{\sigma_i\left(\zeta u^{(n-\varrho+1)/q_i}\right)}.
\end{aligned}$$

Hence

$$\bigoplus_i \bigoplus_{\zeta^{q_i}=1} \mathscr{E}^{\sigma_i\left(\zeta u^{(n-\varrho+1)/q_i}\right)} \otimes \left(\rho^+ P_i^{1/q_i}\right) \cong \bigoplus_{\xi^{n-\varrho+1}=1} \mathscr{E}^{\alpha_{n+1}(\xi u)} \otimes T_{\alpha_{n+1}(\xi u)}$$

The numbers of direct summands in both representations have to be equal. Thus $\sum q_i = n - \varrho + 1$. Moreover, for certain ζ, ξ, i,

$$\mathscr{E}^{\alpha_{n+1}(\xi u)} \cong \mathscr{E}^{\sigma_i\left(\zeta u^{(n-\varrho+1)/q_i}\right)}.$$

The connections $\mathscr{E}^{\alpha_{n+1}(\xi u)}$ are isomorphic for all ξ with $\xi^{n-\varrho+1} = 1$. But this means that all the connections $\mathscr{E}^{\sigma_i\left(\zeta u^{(n-\varrho+1)/q_i}\right)}$ are isomorphic. Hence the sum $\bigoplus_i \mathrm{El}(\eta_i, \sigma_i, P_i)$ only consists of one summand $\mathrm{El}(\eta, \sigma, P)$ due to the refined theorem of Levelt-Turrittin-Malgrange (which says that $\eta_i \mathscr{E}^{\sigma_i} \not\cong \eta_j \mathscr{E}^{\sigma_j}$ if $i \neq j$). Moreover $\alpha_{n+1} = \sigma$ and $\rho = \eta$. This completes the proof. □

4.2.9 Theorem: Let $\varphi(x) = \sum_{i=1}^{n} a_i x^{-i}$ resp. $\psi(y) = \sum_{i=1}^{n} b_i y^{-i}$ be meromorphic functions of pole order n and with $\varrho := \max_i \{(-1)^i a_i + b_i \neq 0\} < n$. Then

(I) After applying the direct image γ_+ with $\gamma : (x, y) \mapsto (x, x+y) =: (u, z)$ to the set $C = \{(x, y) | \varphi'(x) - \psi'(y) = 0\}$, the resulting set C' (see page 53) has $n+1$ parametrizations of the form

$$(u, \tilde{\rho}_k(u)) := \left(u, \sum_{p=1}^{\infty} c_{p,k} u^p\right), \quad k \in \{1, ..., n\}, \text{ and}$$

$$(u, \tilde{\rho}_{n+1}(u)) := \left(u, \sum_{p=1}^{\infty} c_{p,n+1} u^{n-\varrho+p}\right).$$

(II) For all $k \in \{1, ..., n+1\}$, we define

$$\varphi(u) + \psi(\tilde{\rho}_k(u) - u) = \sum_{l=1}^{q_k} d_{l,k} u^{-l} + \sum_{l=0}^{\infty} \tilde{d}_{l,k} u^l =: \tilde{\alpha}_k(u) + \sum_{l=0}^{\infty} \tilde{d}_{l,k} u^l$$

with $q_k = n$ for $k \leq n$, $q_k = \varrho$ for $k = n+1$.

(III) Let $\lambda_k(u) \in u\mathbb{C}[\![u]\!]$ with $\lambda_k'(0) \neq 0$ and $(\tilde{\rho}_k \circ \lambda_k)(u) = u$ for $k \in \{1, ..., n\}$, and $\rho_{n+1}(u) := (\tilde{\rho}_{n+1} \circ \lambda_{n+1})(u) = u^{n-\varrho+1}$. Then for all $k \in \{1, ..., n+1\}$, we set

$$\varphi(\lambda_k(u)) + \psi((\tilde{\rho}_k \circ \lambda_k)(u) - \lambda_k(u)) =: \alpha_k(u) + \delta_k(u),$$

where α_k resp. δ_k denotes the meromorphic resp. holomorphic part. Next we cancel redundant pairs $(\mathrm{Id}, \alpha_k + \delta_k)$ for $k \in \{1, ..., n\}$ as described on page 73. Then, for a suitable renumbering, we receive a minimal set of parametrizations

$$\{(\mathrm{Id}, \alpha_k) | k = 1, ..., r^*, r^* \leq n\} \cup \{(\rho_{n+1}, \alpha_{n+1})\}.$$

Finally we consider the decomposition of the set $\Gamma = \biguplus_{k=1}^{h+1} \Gamma_k$ which we defined on page 73.

It follows (after a suitable renumbering)

$$(\mathrm{El}(\mathrm{Id}, \varphi, R) *_{+_0} \mathrm{El}(\mathrm{Id}, \psi, \mathbf{1}))_0^{\wedge} \cong \left(\bigoplus_{k=1}^{h} \mathrm{El}(\mathrm{Id}, \alpha_k, T_{\alpha_k})\right) \oplus \mathrm{El}(\rho_{n+1}, \alpha_{n+1}, T_{\alpha_{n+1}})$$

with regular connections T_{α_k} for all k. $\square$

4.2.10 Example: Let

$$\varphi(x) = \frac{1}{x} - \frac{1}{2x^2} \text{ and } \psi(y) = \frac{1}{2y^2}.$$

The application of γ_+ with $\gamma : (x, y) \mapsto (x, x + y) =: (u, z)$ to $C = \{(x, y) | \varphi'(x) - \psi'(y) = 0\}$ defines the set C' (see page 53) which we have to parameterize now. We have

$$\varphi'(x) - \psi'(y) = -\frac{1}{x^2} + \frac{1}{x^3} + \frac{1}{y^3} = 0.$$

Applying γ_+, we have the equation

$$-\frac{1}{u^2} + \frac{1}{u^3} + \frac{1}{(z-u)^3} = 0 \Leftrightarrow u^3 + (1-u)(z-u)^3 = 0.$$

So the possible slopes are -1 and -2.
Using the slope -1, the results are $z_k(u) = \tilde{\rho}_k(u) = \left(-\zeta_3^k + 1\right) u + ...,\ k \in \{1, 2\}$.
With the slope -2, we get $z_3(u) = \tilde{\rho}_3(u) = -\frac{1}{3}u^2 + ...$
For the cases $k \in \{1, 2\}$, the computation is the same as in the case $\varrho = n$ (see pages 63 ff.). So here, we will concentrate on $k = 3$. We have

$$\tilde{\alpha}_3(u) = \varphi(u) + \psi(\tilde{\rho}_3(u) - u) = \frac{1}{u} - \frac{1}{2u^2} + \frac{1}{2(\tilde{\rho}_3(u) - u)^2} \mod \mathbb{C}[\![u]\!].$$

In order to check the regular part with respect to $\tilde{\alpha}_3(\eta)$, we have to compute

$$\Psi_\eta \left(\mathscr{E}^{\varphi(u)+\psi(\tilde{\rho}_3(\eta)-u)-\tilde{\alpha}_3(\eta)} \otimes (R \boxtimes S)\right).$$

To do so, let us consider

$$\begin{aligned}
\varphi(u) + \psi\left(\tilde{\rho}_3(\eta) - u\right) - \tilde{\alpha}_3(\eta) &= \frac{1}{u} - \frac{1}{2u^2} + \frac{1}{2(\tilde{\rho}_3(\eta) - u)^2} - \frac{1}{\eta} + \frac{1}{2\eta^2} - \frac{1}{2(\tilde{\rho}_3(\eta) - \eta)^2} \\
&= \frac{1}{u^2\eta^2}\left(u\eta(\eta - u) + \frac{1}{2}\left(u^2 - \eta^2\right)\right) \\
&+ \frac{1}{u^2\eta^2}\left(\frac{1}{2}u^2\eta^2\left((\rho_3(\eta) - u)^{-2} - (\tilde{\rho}_3(\eta) - \eta)^{-2}\right)\right).
\end{aligned}$$

In order to solve the singularity, we need one blowup of the form $(v, \eta) \mapsto (v\eta, \eta)$.

We get

$$
\begin{aligned}
&\frac{1}{v^2\eta^4}\left(v\eta^3(1-v)+\frac{1}{2}\eta^2\left(v^2-1\right)+\frac{1}{2}v^2\eta^4\left(\left(\tilde{\rho}_3(\eta)-v\eta\right)^{-2}-\left(\tilde{\rho}_3(\eta)-\eta\right)^{-2}\right)\right)\\
=&\frac{1}{v^2\eta^2}\left(v\eta(1-v)+\frac{1}{2}\left(v^2-1\right)+\frac{1}{2}v^2\left(\left(\frac{\tilde{\rho}_3(\eta)}{\eta}-v\right)^{-2}-\left(\frac{\tilde{\rho}_3(\eta)}{\eta}-1\right)^{-2}\right)\right)\\
=:&\frac{f_3(v,\eta)}{v^2\eta^2}.
\end{aligned}
$$

Now it is easy to show that

$$
f_3(1,\eta)=0,\ \frac{\partial}{\partial v}f_3(1,\eta)=0,\ \frac{\partial^2}{\partial v^2}f_3(1,\eta)\neq 0.
$$

Note that

$$
\frac{\partial}{\partial v}f_3(v,\eta)=\eta-2v\eta+v+v\left(\frac{\tilde{\rho}_3(\eta)}{\eta}-v\right)^{-2}-v\left(\frac{\tilde{\rho}_3(\eta)}{\eta}-1\right)^{-2}+v^2\left(\frac{\tilde{\rho}_3(\eta)}{\eta}-v\right)^{-3}.
$$

So

$$
\frac{\partial}{\partial v}f_3(1,\eta)=1-\eta+\left(\frac{\tilde{\rho}_3(\eta)}{\eta}-1\right)^{-3}=0\Leftrightarrow\eta^3+(1-\eta)\left(\tilde{\rho}_3(\eta)-\eta\right)^3=0.
$$

But the last equation holds because it fulfills the equation that defines C' - and the pair $(\eta,\tilde{\rho}_3(\eta))$ is one of its solutions. Hence $(v-1)^2$ divides $f_3(v,\eta)$ and so the module $\mathrm{El}\left(\tilde{\rho}_3(u),\tilde{\alpha}_3(u),T_{\tilde{\alpha}_3}\right)$ shows up in the decomposition of the local formal convolution $\left(\mathrm{El}\left(\mathrm{Id},\varphi,R\right)*_{+_0}\mathrm{El}\left(\mathrm{Id},\psi,\mathbf{1}\right)\right)_0^{\wedge}$.

Appendix

A. An alternative formula for the additive convolution

Let us consider an elementary connection $\mathrm{El}(\rho, \varphi, R)$ with $\rho(u) \in u\mathbb{C}[\![u]\!]$, $\varphi(u) \in \mathbb{C}(\!(u)\!)$ and a finite dimensional $\mathbb{C}(\!(u)\!)$-vector space R with regular connection ∇. This connection is defined in a neighbourhood of the irregular singularity located at the origin. Moreover we know from Remark 3.2 that, up to isomorphism, $\mathrm{El}(\rho, \varphi, R)$ is defined over $\mathbb{C}(\{t\}) := \mathbb{C}\{t\}[t^{-1}]$ and we can assume $\varphi(u) \in u^{-1}\mathbb{C}[u^{-1}]$.
Additionally (cf. [1, Proof of Lemma 2.6]), after applying $\lambda(u) \in u\mathbb{C}[\![u]\!]$ with $\lambda'(0) \neq 0$ such that $\mathrm{El}(\rho, \varphi, R) \cong \mathrm{El}(\rho \circ \lambda, \varphi \circ \lambda, R) = \mathrm{El}([u \mapsto u^p], \varphi \circ \lambda, R) =: \mathrm{El}(\tilde{\rho}, \tilde{\varphi}, R)$, we can choose a finite dimensional $\mathbb{C}(\!(t)\!)$-vector space $R^{1/p}$ with $\tilde{\rho}^+ R^{1/p} = R$ and

$$\mathrm{El}(\tilde{\rho}, \tilde{\varphi}, R) = \tilde{\rho}_+ \left(\mathscr{E}^{\tilde{\varphi}} \otimes R\right) = \tilde{\rho}_+ \left(\mathscr{E}^{\tilde{\varphi}} \otimes \tilde{\rho}^+ R^{1/p}\right) = \tilde{\rho}_+ \mathscr{E}^{\tilde{\varphi}} \otimes R^{1/p},$$

where the last equation follows from the projection formula [1, p. 3]. Hence we have the representation $\mathrm{El}(\rho, \varphi, R) \cong \tilde{\rho}_+ \mathscr{E}^{\tilde{\varphi}} \otimes R^{1/p}$ which is a canonical extension of $\mathrm{El}(\rho, \varphi, R)$ to $\mathbb{P}^1$ with regular singularity at ∞. Note that this extension, also called *Katz extension*, is not unique because $R^{1/p}$ is not uniquely defined (cf. [1, Rem. 3.2]). For the general definition of Katz extensions, we refer to the Sections 2.1, 2.2 and 2.4 of Nicholas M. Katz' paper "*On the calculation of some differential galois groups*"[17]. For us, it is enough to know the following result ([18, p. 593]):

A.1 Remark: [17, Th. (2.4.10)] guarantees the existence of a canonical (but not unique) functorial smooth extension $\mathcal{M}$ of a connection M on $k(\!(t)\!)$ to $\mathbb{G}_m = \mathrm{Spec}\,([t, t^{-1}])$ such that $\mathcal{M}$ has a regular singular point at $t = \infty$.

Now let us return to our elementary connection $\mathrm{El}(\rho, \varphi, R)$ from above. This vector space just depends on its behaviour around the singularity. So this observation gives rise to the question whether formal germs of the Fourier transform $F_\pm(\mathcal{M})$ or the formal germs of the additive convolution $\mathcal{M} *_{+_0} \mathcal{N}$ of meromorphic connections $\mathcal{M}, \mathcal{N}$ can also be expressed in terms of the formal germs defined by $\mathcal{M}, \mathcal{N}$ at their singularities (cf. [2, Introduction]).

With the help of of a decomposition formula for the local formal Fourier transform (Section A.1) and the expression of the additive convolution through Fourier transforms (Section A.2), we are prepared to give the main statement - an alternative decomposition formula for the additive convolution of meromorphic connections (unfortunately this is a purely formal computation, i.e., one cannot use the obtained formula for the study of Stokes structures). We conclude the chapter by giving an application for elementary formal meromorphic connections.

A.1. A decomposition formula for the local formal Fourier transform

In his paper "*Microlocalization and stationary phase*"[2], Ricardo G. Lopez proves a statement that is essential for us. But before noting it, we have to give a short review of his definition of the microlocalization of modules over the Weyl algebra $\mathbb{W}_t := \mathbb{C}[t]\langle\partial_t\rangle$. Let M be a holonomic module over $\mathbb{W}_t$. The Fourier antiinvolution is the morphism of $\mathbb{C}$-algebras $\mathbb{W}_t \to \mathbb{W}_\tau$ given by $t \mapsto -\partial_\tau, \partial_t \mapsto \tau$. The Fourier transform of M is defined as the $\mathbb{W}_\tau$-module $F(M) := \mathbb{W}_\tau \otimes_{\mathbb{W}_t} M$, where $\mathbb{W}_\tau$ is regarded as a right $\mathbb{W}_t$-module via the Fourier morphism. For $m \in M$, we put $\hat{m} = 1 \otimes m \in F(M)$.
On $\mathbb{C}\left(\left(\tau^{-1}\right)\right) = \mathbb{C}[\![\tau^{-1}]\!][\tau]$, we consider the derivation $\partial_{\tau^{-1}} = -\tau^2\partial_\tau$. So if V is a $\mathbb{C}\left(\left(\tau^{-1}\right)\right)$-vector space, a connection on V is a $\mathbb{C}$-linear map $\nabla : V \to V$ satisfying the Leibniz rule

$$\nabla(\alpha \cdot v) = \partial_{\tau^{-1}}(\alpha) \cdot v + \alpha \cdot \nabla(v) \text{ for all } \alpha \in \mathbb{C}\left(\left(\tau^{-1}\right)\right),\ v \in V.$$

We set $M_\infty := \mathbb{C}[\![t^{-1}]\!]\langle\partial_t\rangle \otimes_{\mathbb{C}[t^{-1}]\langle\partial_t\rangle} M[t^{-1}]$, and for $c \in \mathbb{C}$, put $M_c := \mathbb{C}[\![t_c]\!]\langle\partial_{t_c}\rangle \otimes_{\mathbb{W}_t} M$, where $t_c = t - c$. Then the (ordinary) microlocalization of M at $c \in \mathbb{C}$ is defined as

$$\mathscr{F}^{(c,\infty)}(M) := \mathscr{F}^{(c,\infty)} \otimes_{\mathbb{W}_t} M$$

with the ring

$$\mathscr{F}^{(c,\infty)} = \bigcup_r \left\{ \sum_{i \geq r} a_i(t_c)\tau^i \,\middle|\, a_i(t_c) \in \mathbb{C}[\![t_c]\!],\ r \in \mathbb{Z} \right\}.$$

$\mathscr{F}^{(c,\infty)}(M)$ has a structure of a $\mathbb{C}\left(\left(\tau^{-1}\right)\right)$-vector space with a connection given by left multiplication by $\tau^2 \cdot (t_c + c) = \tau^2 \cdot t$. We have

$$\mathscr{F}^{(c,\infty)}(M) = \mathscr{F}^{(c,\infty)} \otimes_{\mathbb{C}[\![t_c]\!]\langle\partial_{t_c}\rangle} (\mathbb{C}[\![t_c]\!]\langle\partial_{t_c}\rangle \otimes_{\mathbb{W}_t} M) = \mathscr{F}^{(c,\infty)} \otimes_{\mathbb{C}[\![t_c]\!]\langle\partial_{t_c}\rangle} M_c.$$

Thus $\mathscr{F}^{(c,\infty)}(M)$ only depends on the formal germ M_c.
The (∞, ∞)-microlocalization of M is defined as

$$\mathscr{F}^{(\infty,\infty)}(M) := \mathscr{F}^{(\infty,\infty)} \otimes_{\mathbb{C}[t^{-1}]\langle\partial_t\rangle} M[t^{-1}],$$

where

$$\mathscr{F}^{(\infty,\infty)} = \bigcup_r \left\{ \sum_{i \geq r} a_i\left(t^{-1}\right)\tau^i \,\middle|\, a_i\left(t^{-1}\right) \in \mathbb{C}[\![t^{-1}]\!],\ r \in \mathbb{Z} \right\}$$

defines a ring. $\mathscr{F}^{(\infty,\infty)}(M)$ also has a structure of a $\mathbb{C}\left((\tau^{-1})\right)$-vector space with a connection defined by $\nabla(\alpha \otimes m) := \partial_{\tau^{-1}}(\alpha) \otimes m + \tau^2\alpha \otimes t \cdot m$. We have

$$\mathscr{F}^{(\infty,\infty)}(M) = \mathscr{F}^{(\infty,\infty)} \otimes_{\mathbb{C}[\![t^{-1}]\!]\langle\partial_t\rangle} M_\infty.$$

Thus $\mathscr{F}^{(\infty,\infty)}(M)$ depends only on the formal germ M_∞.
After the following Definition (see [2, Definition, p. 7]), we are ready to give the main statement ([2, Theorem, p.7]) of this section:

A.1.1 Definition: If N is a $\mathbb{W}_\tau$-module, its formal germ at infinity is the $\mathbb{C}\left((\tau^{-1})\right)$-vector space $\mathcal{N}_\infty = \mathbb{C}\left((\tau^{-1})\right) \otimes_{\mathbb{C}[\tau]} N$, which is endowed with the connection

$$\nabla(\alpha \otimes n) = \partial_{\tau^{-1}}(\alpha) \otimes n - \alpha \otimes \tau^2\partial_\tau n.$$

A.1.2 Theorem: Let M be a holonomic $\mathbb{W}_t$-module and $\mathrm{Sing}(M) \subset \mathbb{C}$ the set of singularities of M. Then the map

$$\Upsilon : F(M)_\infty \to \bigoplus_{c \in \mathrm{Sing}(M) \cup \{\infty\}} \mathscr{F}^{(c,\infty)}(M),$$

given by $\Upsilon(\alpha \otimes \hat{m}) = \oplus_c(\alpha \otimes m)$, is an isomorphism of $\mathbb{C}\left((\tau^{-1})\right)$-vector spaces with connection. □

A.1.3 Remark: Of course, we want to work with meromorphic connections, i.e., finite dimensional $\mathbb{C}((t))$-vector spaces with connection, instead of modules over the Weyl algebra which are just holonomic. C. Sabbah gives a construction of $\mathscr{F}^{(0,\infty)}(M)$ for a meromorphic connection (M, ∇) (see [1, Introduction]).

Theorem A.1.2 is the first key result which we need to give the decomposition formula for the additive convolution.

A.2. Expressing the additive convolution through Fourier transforms

Here we have the second essential statement (cf. [19, Key Lemma 12.2.3(5)]).

A.2.1 Lemma: For meromorphic connections M, N, we have

$$M *_+ N \cong F^{-1}\left(F(M) \otimes F(N)\right).$$

□

A.3. Decomposition of the additive convolution in local summands

The goal of this section is the Proof of an alternative convolution formula. We have to differ between the cases $e = \infty$ and $e \in \mathbb{C}$.

Case 1) $e = \infty$: Consider Katz extensions M, N of two meromorphic connections. We have

$$(M *_+ N)^\wedge_\infty \cong \left(F^{-1}(F(M) \otimes F(N))\right)^\wedge_\infty$$

due to Lemma A.2.1. When we denote by $F_+ := F$ the algebraic Laplace transform with kernel $e^{t\tau}$, we are able to define $F_- := F^{-1}$ as the Laplace transform with kernel $e^{-t\tau}$ (see [1, Introduction]). With this notation,

$$(M *_+ N)^\wedge_\infty \cong (F_-(F_+(M) \otimes F_+(N)))^\wedge_\infty .$$

Analogously, we have pairs of inverse transforms $\left(\mathscr{F}_\pm^{(s,\infty)}, \mathscr{F}_\mp^{(\infty,s)}\right)$ for all $s \in \mathbb{C}$ and $\left(\mathscr{F}_\pm^{(\infty,\infty)}, \mathscr{F}_\mp^{(\infty,\infty)}\right)$. So we can apply Theorem A.1.2. This gives

$$(F_-(F_+(M) \otimes F_+(N)))^\wedge_\infty \cong \bigoplus_{f \in \mathrm{Sing}(F_+(M) \otimes F_+(N))} \mathscr{F}_-^{(f,\infty)}\left(F_+(M) \otimes F_+(N)\right).$$

After Section A.1, the expression $\mathscr{F}_-^{(f,\infty)}\left(F_+(M) \otimes F_+(N)\right)$ just depends on the formal germ $\left(F_+(M) \otimes F_+(N)\right)_f$, and we can write

$$\mathscr{F}_-^{(f,\infty)}\left(F_+(M) \otimes F_+(N)\right) = \mathscr{F}_-^{(f,\infty)} \otimes_{\mathbb{C}[\![t_f]\!]\langle \partial_{t_f} \rangle} \left(F_+(M) \otimes F_+(N)\right)_f .$$

But clearly,

$$(F_+(M) \otimes F_+(N))_f = F_+(M)_f \otimes F_+(N)_f.$$

Now the deciding question is whether we can also apply Theorem A.1.2 to $F_+(M)_f$ in the case $f \neq \infty$, i.e., whether the map

$$\begin{aligned} \Upsilon_f : F_+(M)_f &\to \bigoplus_{c \in \mathrm{Sing}(M)} \mathscr{F}_+^{(c,f)}(M) \\ \alpha \otimes \hat{m} &\mapsto \oplus_c(\alpha \otimes m) \end{aligned}$$

is an isomorphism of $\mathbb{C}((\tau^{-1}))$-vector spaces with connection. The answer is yes. In order to show this, let us first consider the diagram

$$\begin{array}{ccccccccccccc} M & & M_0 & & \bigoplus_{s\neq 0} M_s & & M_\infty^{>1} & & M_\infty^{=1} & & M_\infty^{<1} & & M \\ F_- \downarrow & \mathscr{F}_-^{(0,\infty)} & \downarrow\uparrow & \oplus\mathscr{F}_-^{(s,\infty)} \;\; \mathscr{F}_+^{(\infty,0)} & \downarrow\uparrow & \mathscr{F}_-^{(\infty,\infty)} \;\; \oplus\mathscr{F}_+^{(\infty,s)} & \downarrow\uparrow & \oplus\mathscr{F}_-^{(\infty,\hat{s})} \;\; \mathscr{F}_+^{(\infty,\infty)} & \downarrow\uparrow & \mathscr{F}_-^{(\infty,0)} \;\; \oplus\mathscr{F}_+^{(\hat{s},\infty)} & \downarrow\uparrow & \mathscr{F}_+^{(0,\infty)} & \uparrow F_+ \\ \hat{M} & & \hat{M}_\infty^{<1} & & \hat{M}_\infty^{=1} & & \hat{M}_\infty^{>1} & & \bigoplus_{\hat{s}\neq 0} \hat{M}_s & & \hat{M}_0 & & \hat{M} \end{array}$$

(where $\hat{M}$ denotes the formalization of M and $s, \hat{s} \neq \infty$ the singularities of $M, \hat{M}$) which decomposes M with respect to the slopes at infinity (see [1, Introduction]). Note that $M = \hat{M}$ as M already is a Katz extension. We just use $\hat{M}$ in the diagram for a better discriminability. Now we can deduce from this diagram that $\mathscr{F}_\pm^{(a,b)}(M)$ only has a non-trivial effect in the case $a = \infty \vee b = \infty$. In the case $a, b \neq \infty$, we have $\mathscr{F}_\pm^{(a,b)}(M) = 0$. Moreover

$$\begin{aligned} F_+(M) \cong \mathscr{F}_+^{(\infty,0)}\left(M_\infty^{<1}\right) &\oplus \left(\bigoplus \mathscr{F}_+^{(\infty,s)}\left(M_\infty^{=1}\right)\right) \oplus \mathscr{F}_+^{(\infty,\infty)}\left(M_\infty^{>1}\right) \\ &\oplus \left(\bigoplus \mathscr{F}_+^{(\hat{s},\infty)}\left(M_{\hat{s}}\right)\right) \oplus \mathscr{F}_+^{(0,\infty)}\left(M_0\right) \end{aligned}$$

implies $F_+(M)_f \cong \mathscr{F}_+^{(\infty,f)}(M)$. But as $\mathscr{F}_\pm^{(a,b)}(M) = 0$ for $a, b \neq \infty$, this is exactly the statement of Theorem A.1.2 applied to $f \neq \infty$. This means,

$$F_+(M)_f \cong \mathscr{F}_+^{(\infty,f)}(M) \cong \bigoplus_{c \in \mathrm{Sing}(M)} \mathscr{F}_+^{(c,f)}(M).$$

Hence Υ_f is an isomorphism. The same holds for $F_+(N)_f$, of course.
Additionally we have shown that both $F_+(M)_f$ and $F_+(N)_f$ depend on the singularities of M and N for all $f \in \mathbb{P}^1$. So the singularities of both $F_+(M)$ and $F_+(N)$ depend on the singularities of M and N. Now the singularities of $F_+(M) \otimes F_+(N)$ depend on the

singularities of M and N, i.e.,

$$\mathrm{Sing}(F_+(M)\otimes F_+(N))\cup\{\infty\} = \mathrm{Sing}(F_+(M)\otimes F_+(N)) \subseteq \mathrm{Sing}(M)\cup \mathrm{Sing}(N)$$

because the divisor D can be written in terms of the poles of M and N [20, p. 4]. Hence

$$\begin{aligned}
(M *_+ N)^\wedge_\infty &\cong (F_-(F_+(M)\otimes F_+(N)))^\wedge_\infty \\
&\cong \bigoplus_{f\in\mathrm{Sing}(F_+(M)\otimes F_+(N))} \mathscr{F}_-^{(f,\infty)}\,(F_+(M)\otimes F_+(N)) \\
&\cong \bigoplus_f \mathscr{F}_-^{(f,\infty)} \otimes_{\mathbb{C}[\![t_f]\!]\langle\partial_{t_f}\rangle} (F_+(M)\otimes F_+(N))_f \\
&= \bigoplus_f \mathscr{F}_-^{(f,\infty)} \otimes (F_+(M)_f\otimes F_+(N)_f) \\
&\cong \bigoplus_f \mathscr{F}_-^{(f,\infty)} \otimes \left(\left(\bigoplus_{c\in\mathrm{Sing}(M)} \mathscr{F}_+^{(c,f)}(M)\right)\otimes F_+(N)_f\right) \\
&\cong \bigoplus_f \mathscr{F}_-^{(f,\infty)} \otimes \left(\left(\bigoplus_c \mathscr{F}_+^{(c,f)}(M)\right)\otimes \left(\bigoplus_{d\in\mathrm{Sing}(N)} \mathscr{F}_+^{(d,f)}(N)\right)\right) \\
&\cong \bigoplus_{f\neq\infty} \mathscr{F}_-^{(f,\infty)} \left(\left(\mathscr{F}_+^{(\infty,f)}(M)\right)\otimes\left(\mathscr{F}_+^{(\infty,f)}(N)\right)\right) \\
&\qquad \oplus \mathscr{F}_-^{(\infty,\infty)}\left(\left(\bigoplus_c \mathscr{F}_+^{(c,\infty)}(M)\right)\otimes\left(\bigoplus_d \mathscr{F}_+^{(d,\infty)}(N)\right)\right).
\end{aligned}$$

<u>Case 2) $e\neq\infty$:</u> We have shown above that we can also use Theorem A.1.2 in the case $e\neq\infty$. Reapplication of this statement and the fact $\mathscr{F}_-^{(f,e)}(\_)=0$ for $f,e\neq\infty$ gives

$$\begin{aligned}
(M *_+ N)^\wedge_e &\cong \bigoplus_{f\in\mathrm{Sing}(F_+(M)\otimes F_+(N))} \mathscr{F}_-^{(f,e)}(F_+(M)\otimes F_+(N)) \\
&\cong \mathscr{F}_-^{(\infty,e)} \otimes (F_+(M)\otimes F_+(N))_\infty = \mathscr{F}_-^{(\infty,e)}\otimes(F_+(M)_\infty\otimes F_+(N)_\infty) \\
&\cong \mathscr{F}_-^{(\infty,e)} \otimes \left(\left(\bigoplus_{c\in\mathrm{Sing}(M)} \mathscr{F}_+^{(c,\infty)}(M)\right)\otimes F_+(N)_\infty\right) \\
&\cong \mathscr{F}_-^{(\infty,e)} \otimes \left(\left(\bigoplus_c \mathscr{F}_+^{(c,\infty)}(M)\right)\otimes\left(\bigoplus_{d\in\mathrm{Sing}(N)} \mathscr{F}_+^{(d,\infty)}(N)\right)\right) \\
&= \mathscr{F}_-^{(\infty,e)} \left(\left(\bigoplus_c \mathscr{F}_+^{(c,\infty)}(M)\right)\otimes\left(\bigoplus_d \mathscr{F}_+^{(d,\infty)}(N)\right)\right).
\end{aligned}$$

As the Fourier tranform, particularly the local Fourier transforms, have a right adjoint, they respect direct sums. □

A.3.1 Remark: Looking at both cases, we can see that e is important for the computation of $M *_+ N$, of course. But still, the formulas just depend on the local types of M, N at their singularities.

A.4. The special case of elementary formal meromorphic connections

Of course, this convolution formula also holds in the case that M and N are elementary formal meromorphic connections. So let

$$M = \mathrm{El}(\rho(x), \varphi(x), R) \text{ and } N = \mathrm{El}(\eta(y), \psi(y), S)$$

with $\rho(x) = x^p$, $\eta(y) = y^q$, $\varphi(x) = \sum_{l=1}^{n} a_l x^{-l}$, $\psi(y) = \sum_{j=1}^{m} b_j y^{-j}$, R (resp. S) a $\mathbb{C}((x))$- (resp. $\mathbb{C}((y))$-)vector space of rank r (resp. s) with regular connection. As in the previous chapters, we assume that R and S both have connections with regular singularities at 0 and ∞ (and no other pole). This means, $S := \mathrm{Sing}(M) = \mathrm{Sing}(N) = \{0, \infty\}$. Hence

$$\begin{aligned}&(\mathrm{El}(\rho,\varphi,R) *_+ \mathrm{El}(\eta,\psi,S))^\wedge_e \\ \cong &\begin{cases} \bigoplus_{c,d,f\in S} \mathscr{F}_-^{(f,\infty)}\left(\mathscr{F}_+^{(c,f)}(\mathrm{El}(\rho,\varphi,R)) \otimes \mathscr{F}_+^{(d,f)}(\mathrm{El}(\eta,\psi,S))\right), & \text{if } e = \infty, \\ \bigoplus_{c,d\in S} \mathscr{F}_-^{(\infty,e)}\left(\mathscr{F}_+^{(c,\infty)}(\mathrm{El}(\rho,\varphi,R)) \otimes \mathscr{F}_+^{(d,\infty)}(\mathrm{El}(\eta,\psi,S))\right), & \text{else.}\end{cases}\end{aligned}$$

So the decomposition in local summands gives an alternative formula for the computation of the local formal additive convolution.

A.4.1 Remark: We have

$$(\mathrm{El}(\rho,\varphi,R) *_+ \mathrm{El}(\eta,\psi,S))^\wedge_e \cong (\mathrm{El}(\rho,\varphi,R) *_{+_0} \mathrm{El}(\eta,\psi,S))^\wedge_e$$

because after [6, Example 0.4], $\mathrm{El}(\rho,\varphi,R) *_{+_0} \mathrm{El}(\eta,\psi,S)$ is the only cohomology module with not just punctual support.

Now let us start with the decomposition of $(\mathrm{El}(\rho,\varphi,R) *_+ \mathrm{El}(\eta,\psi,S))^\wedge_e$ with $e \neq 0$. We already know that

$$\begin{aligned}
\mathscr{F}_\pm^{(0,\infty)} &: (\mathbb{C}((t)))^{pr} \to (\mathbb{C}((\tau)))^{(p+n)r}, \mathrm{El}(\rho,\varphi,R) \mapsto \mathrm{El}\left(\mp\frac{\rho'}{\varphi'}, \varphi - \frac{\rho}{\rho'}\varphi', R\otimes L_n\right),\\
\mathscr{F}_\pm^{(\infty,0)} &: (\mathbb{C}((\tau)))^{pr} \to (\mathbb{C}((t)))^{(p-n)r}, \mathrm{El}(\rho,\varphi,R) \mapsto \mathrm{El}\left(\pm\frac{\rho^2\varphi'}{\rho'}, \varphi + \frac{\rho}{\rho'}\varphi', R\otimes L_n\right),\\
\mathscr{F}_\pm^{(s,\infty)} &: (\mathbb{C}((t)))^{pr} \to (\mathbb{C}((\tau)))^{(p+n)r}, \mathrm{El}(\rho,\varphi,R) \mapsto \mathscr{E}^{\pm\frac{s}{\tau}} \otimes \mathscr{F}_\pm^{(0,\infty)}\mathrm{El}\left(\rho,\varphi,R\right),\\
\mathscr{F}_\pm^{(s,\infty)} &: (\mathbb{C}((t)))^{r} \to (\mathbb{C}((\tau)))^{r}, \qquad\qquad R \mapsto \mathrm{El}\left(\mathrm{Id}, \pm\frac{c}{\tau}, \mathscr{F}_\pm^{(0,\infty)}R\right),\\
\mathscr{F}_\pm^{(\infty,\infty)} &: (\mathbb{C}((t)))^{pr} \to (\mathbb{C}((\tau)))^{(n-p)r}, \mathrm{El}(\rho,\varphi,R) \mapsto \mathrm{El}\left(\pm\frac{\rho'}{\varphi'\rho^2}, \varphi + \frac{\rho}{\rho'}\varphi', R\otimes L_n\right),
\end{aligned}$$

are transforms of finite dimensional vector spaces with connection due to [1, Th. 5.1], [1, Rem. 5.3(4)], and [1, Sect. 5.c]. But the diagram on page 84 shows that $\mathscr{F}^{(\infty,0)}$ can only be applied to connections with slope < 1 at infinity, $\mathscr{F}^{(\infty,\infty)}$ can only be applied to connections with slope > 1 at infinity, and $\mathscr{F}^{(\infty,a)}$ $(a \in \mathbb{C}^\times)$ can only be applied to connections with slope $= 1$ at infinity. However the given elementary formal meromorphic connections have a regular singularity at infinity. This means

$$\begin{aligned}
\mathscr{F}_\pm^{(\infty,0)}\mathrm{El}(\rho,\varphi,R) &= \mathscr{F}_\pm^{(\infty,0)}\mathrm{El}(\eta,\psi,S) = 0,\\
\mathscr{F}_\pm^{(\infty,\infty)}\mathrm{El}(\rho,\varphi,R) &= \mathscr{F}_\pm^{(\infty,\infty)}\mathrm{El}(\eta,\psi,S) = 0,\\
\mathscr{F}_-^{(\infty,e)}\left(\mathscr{F}_+^{(0,\infty)}(\mathrm{El}(\rho,\varphi,R)) \otimes \mathscr{F}_+^{(0,\infty)}(\mathrm{El}(\eta,\psi,S))\right) &= 0,\ e\neq 0.
\end{aligned}$$

This already gives the following result.

A.4.2 Lemma: Let

$$\mathrm{El}(\rho,\varphi,R) \text{ and } \mathrm{El}(\eta,\psi,S)$$

be elementary formal meromorphic connections with irregular singularity at zero and regular singularity at infinity. Then for $e \in \mathbb{C}^\times$ and $e = \infty$, we have

$$(\mathrm{El}(\rho,\varphi,R) *_+ \mathrm{El}(\eta,\psi,S))^\wedge_e = 0.$$

□

Next we want to compute a decomposition of $(\mathrm{El}(\rho,\varphi,R) *_+ \mathrm{El}(\eta,\psi,S))^\wedge_0$. For this, we need the formula for the tensor product of elementary formal meromorphic connections (see [1, Prop. 3.8]):

A.4.3 Proposition: Given connections $\mathrm{El}([u \mapsto u^{p_1}], \varphi_1, R_1)$ and $\mathrm{El}([v \mapsto v^{p_2}], \varphi_2, R_2)$, we set $d = \gcd(p_1, p_2), p_i' = p_i/d$, and $\tilde{\rho}_i(\omega) = \omega^{p_i'}$. Then we define

$$\begin{aligned}
\rho(\omega) &= \omega^{p_1 p_2/d}, \\
\varphi^{(k)} &= \varphi_1\left(\omega^{p_2'}\right) + \varphi_2\left(\left[e^{2\pi i k d/p_1 p_2}\omega\right]^{p_1'}\right) \quad (k = 0, ..., d-1), \\
R &= \tilde{\rho}_2^+ R_1 \otimes \tilde{\rho}_1^+ R_2.
\end{aligned}$$

With this notation,

$$\mathrm{El}\left([u \mapsto u^{p_1}], \varphi_1, R_1\right) \otimes \mathrm{El}\left([v \mapsto v^{p_2}], \varphi_2, R_2\right) \cong \bigoplus_{k=0}^{d-1} \mathrm{El}\left(\rho, \varphi^{(k)}, R\right).$$

□

Now, in order to compute $(\mathrm{El}(\rho, \varphi, R) *_+ \mathrm{El}(\eta, \psi, S))_0^\wedge$, one only needs the formulas for the local formal Laplace transforms listed on the last page and Proposition A.4.3. Unfortunately there is a little obstacle: Proposition A.4.3 can only be applied to elementary connections $\mathrm{El}(\rho, \varphi, R)$ and $\mathrm{El}(\eta, \psi, S)$ if the ramifications ρ and η have the canonical forms $\rho(x) = x^{p_1}$ and $\eta(y) = y^{p_2}$. So the computation of the convolution may require exhausting computations.

A.4.4 Example: Let us compute

$$\left(\mathrm{El}\left(\left[x \mapsto \frac{1}{2}x^2\right], \frac{1}{3x^3}, R\right) *_+ \mathrm{El}\left(\mathrm{Id}, \frac{1}{2y^2}, S\right)\right)_0^\wedge.$$

The computations of this section show that this convolution problem reduces to the computation of

$$\mathscr{F}_-^{(\infty,0)}\left(\mathscr{F}_+^{(0,\infty)}\left(\mathrm{El}\left(\left[x \mapsto \frac{1}{2}x^2\right], \frac{1}{3x^3}, R\right)\right) \otimes \mathscr{F}_+^{(0,\infty)}\left(\mathrm{El}\left(\mathrm{Id}, \frac{1}{2y^2}, S\right)\right)\right).$$

We have

$$\begin{aligned}
\mathscr{F}_+^{(0,\infty)}\left(\mathrm{El}\left(\left[x \mapsto \frac{1}{2}x^2\right], \frac{1}{3x^3}, R\right)\right) &\cong \mathrm{El}\left(\left[u \mapsto u^5\right], \frac{5}{6u^3}, R \otimes L_3\right) \\
\mathscr{F}_+^{(0,\infty)}\left(\mathrm{El}\left(\mathrm{Id}, \frac{1}{2y^2}, S\right)\right) &\cong \mathrm{El}\left(\left[v \mapsto v^3\right], \frac{3}{2v^2}, S \otimes L_2\right)
\end{aligned}$$

with the new variables u, v. Now, as $d = \gcd(3,5) = 1$,

$$\mathscr{F}_{+}^{(0,\infty)}\left(\mathrm{El}\left(\left[x \mapsto \frac{1}{2}x^2\right], \frac{1}{3x^3}, R\right)\right) \otimes \mathscr{F}_{+}^{(0,\infty)}\left(\mathrm{El}\left(\mathrm{Id}, \frac{1}{2y^2}, S\right)\right)$$
$$\cong \mathrm{El}\left(\left[\omega \mapsto \omega^{15}\right], \frac{5}{6\omega^9} + \frac{3}{2\omega^{10}}, \tilde{\rho}_2^+\left(R \otimes L_3\right) \otimes \tilde{\rho}_1^+\left(S \otimes L_2\right)\right)$$

with $\tilde{\rho}_k$ defined as in Proposition A.4.3. Finally

$$\mathscr{F}_{-}^{(\infty,0)}\left(\mathrm{El}\left(\left[\omega \mapsto \omega^{15}\right], \frac{5}{6\omega^9} + \frac{3}{2\omega^{10}}, \tilde{\rho}_2^+\left(R \otimes L_3\right) \otimes \tilde{\rho}_1^+\left(S \otimes L_2\right)\right)\right)$$
$$\cong \mathrm{El}\left(\left[\tau \mapsto \tau^5 + \frac{1}{2}\tau^6\right], \frac{1}{3\tau^9} + \frac{1}{2\tau^{10}}, \left(\tilde{\rho}_2^+\left(R \otimes L_3\right) \otimes \tilde{\rho}_1^+\left(S \otimes L_2\right)\right) \otimes L_{10}\right).$$

with the new variable τ. Due to $T \otimes L_{2k} \cong T$ for all $k \in \mathbb{N}$ (see [1, Remark 5.3(3)], we get the final result

$$\left(\mathrm{El}\left(\left[x \mapsto \frac{1}{2}x^2\right], \frac{1}{3x^3}, R\right) *_+ \mathrm{El}\left(\mathrm{Id}, \frac{1}{2y^2}, S\right)\right)_0^{\wedge}$$
$$\cong \mathrm{El}\left(\left[\tau \mapsto \tau^5 + \frac{1}{2}\tau^6\right], \frac{1}{3\tau^9} + \frac{1}{2\tau^{10}}, \tilde{\rho}_2^+\left(R \otimes L_3\right) \otimes \tilde{\rho}_1^+ S\right).$$

Bibliography

[1] C. Sabbah. *An explicit stationary phase formula for local formal Fourier-Laplace transform*, Contemporary Mathematics, 2007;
http://arxiv.org/abs/0706.3570

[2] R. G. Lopez. *Microlocalization and stationary phase*, Asian Journal of Mathematics, Volume 8, Number 4 (2004), 747-768;
http://arxiv.org/abs/math/0307366

[3] M. Kashiwara. *Vanishing cycle sheaves and holonomic systems of differential equations*, Algebraic geometry, Lecture Notes in Math., vol. 1016, Springer, Berlin, 1983;

[4] C. Sabbah. *$\mathscr{D}_X$-modules et cycles évanescents.*
In: Z. Mebkhout. *Le formalisme des six opérations de Grothendieck pour les $\mathscr{D}_X$-modules cohérents*, Travaux en cours 46, pages 201-239, Hermann, Paris, 1988;

[5] Y. Laurent, B. Malgrange. *Cycles proches, spécialisation et $\mathscr{D}$-modules*, Ann. Inst. Fourier t.45, n. 5, p. 1353-1405, 1995;

[6] C. Roucairol. *Formal structure of direct image of holonomic $\mathscr{D}$-modules of exponential type*, manuscripta mathematica 124, pages 299-318, 2007;
http://arxiv.org/abs/math/0604134

[7] C. Roucairol. *Formal structure of direct image of some $\mathscr{D}$ modules*, Publ. RIMS 42 (4), Kyoto Univ. (2006);
http://math.univ-angers.fr/~roucairo/formal.pdf

[8] Claude Sabbah. *Introduction to Stokes structures*, Lecture Notes In Mathematics 2060, Springer, 2013;

[9] J.-B. Teyssier. *Autour de l'irrégularité des connexions méromorphes*, PhD thesis, 2013;
http://jbteyssier.com/papers/jbteyssier_these.pdf

[10] C. Sabbah, Philippe Maisonobe. *Aspects of the theory of $\mathscr{D}$-modules*, Kaiserslautern 2002, revised version July 2011;
http://www.math.polytechnique.fr/cmat/sabbah/livres.html

[11] S. Arkhipov, N. Nikolaev. *Topics in algebraic geometry: $\mathscr{D}$-Modules*, Lecture Notes, 2012;
http://wiki.math.toronto.edu/TorontoMathWiki/images/c/ce/MAT1191_Lecture_Notes.pdf

[12] R. Hotta, K. Takeuchi, T. Tanisaki. *$\mathscr{D}$-modules, perverse sheaves, and representation theory*, Progress in Mathematics 236, Birkhäuser, 2008; www.math.columbia.edu/ scautis/dmodules/hottaetal.pdf

[13] C. Roucairol. *L'irrégularité du complexe* $f_+ (\mathcal{O}_{\mathbb{C}^n} e^g)$, Thesis n. 619, Angers University (2004)

[14] J. R. Sendra, F. Winkler, S. Perez-Diaz, *Rational algebraic curves - A computer algebra approach*, Algorithms and Computation in Mathematics, Springer, 2008;

[15] A. Lipkovski. *Newton polyhedra and Irreducibility*, Mathematische Zeitschrift, Volume 199, Issue 1, pages 119 - 127, Springer, 1988; http://link.springer.com/article/10.1007%2FBF01160215

[16] M. Iwami. *Extension of expansion base algorithm for multivariate analytic factorization including the case of singular leading coefficient*, ACM SIGSAM Bulletin, Vol 39, No. 4, pages 122 - 126, 2005; http://www.sigsam.org/bulletin/articles/154/IW.pdf

[17] N. M. Katz. *On the calculation of some differential galois groups*, Inventiones mathematicae, pages 13 - 61, Springer, 1987; https://web.math.princeton.edu/~nmk/old/calcdiffgal.pdf

[18] S. Bloch, H. Esnault. *Local fourier transforms and rigidity for $\mathscr{D}$-modules*, Asian Journal of Mathematics, Vol. 8, No. 4, pages 587-606, Dec. 2004; projecteuclid.org/euclid.ajm/1118669692

[19] N. M. Katz. *Exponential sums and differential equations*, Annals of Mathematics Studies, Am-124, Princeton University Press, 1990; https://web.math.princeton.edu/~nmk/

[20] B. Pym. *Geometry Retreat 2012*, University of Toronto, Lecture Notes, 2012; http://www.math.toronto.edu/nikolaev/work.html#LectureNotes

Lebenslauf

Persönliche Daten:

Name	Robert Gelb
Geburtsdatum	27.04.1985 in Augsburg

Schulbildung

1995 - 2004	Besuch des Peutinger Gymnasiums in Augsburg Abschluss: Abitur

Wehrdienst

2004 - 2005	Ableistung der Wehrpflicht in Bayreuth und Fürstenfeldbruck

Studium

2005 - 2010	Studium der Mathematik und Volkswirtschaftslehre an der Universität Augsburg Abschluss: Diplom

Berufsweg

2010 - 2014	Wissenschaftlicher Mitarbeiter an der Universität Augsburg
seit 2014	Consultant bei Aon Hewitt

December 22, 2014